Improving Natural Resource Management

Statistics in Practice

Series Advisors

Human and Biological Sciences
Stephen Senn
University of Glasgow, UK

Earth and Environmental Sciences
Marian Scott
University of Glasgow, UK

Industry, Commerce and Finance
Wolfgang Jank
University of Maryland, USA

Statistics in Practice is an important international series of texts which provide detailed coverage of statistical concepts, methods and worked case studies in specific fields of investigation and study.

With sound motivation and many worked practical examples, the books show in down-to-earth terms how to select and use an appropriate range of statistical techniques in a particular practical field within each title's special topic area.

The books provide statistical support for professionals and research workers across a range of employment fields and research environments. Subject areas covered include medicine and pharmaceutics; industry, finance and commerce; public services; the earth and environmental sciences, and so on.

The books also provide support to students studying statistical courses applied to the above areas. The demand for graduates to be equipped for the work environment has led to such courses becoming increasingly prevalent at universities and colleges.

It is our aim to present judiciously chosen and well-written workbooks to meet everyday practical needs. Feedback of views from readers will be most valuable to monitor the success of this aim.

A complete list of titles in this series appears at the end of the volume.

Improving Natural Resource Management

Ecological and Political Models

Timothy C. Haas

Sheldon B. Lubar School of Business
University of Wisconsin-Milwaukee, USA

A John Wiley and Sons, Ltd., Publication

Registered office
John Wiley & Sons Ltd, The Atrium, Southern Gate, Chichester, West Sussex, PO19 8SQ, United Kingdom

For details of our global editorial offices, for customer services and for information about how to apply for permission to reuse the copyright material in this book please see our website at www.wiley.com.

Library of Congress Cataloging-in-Publication Data

Haas, Timothy.
 Improving natural resource management : ecological and political models / Timothy Haas.
   p. cm.
 Includes bibliographical references and index.
 ISBN 978-0-470-66113-0 (cloth)
  1. Ecosystem management. 2. Ecosystem management–Simulation methods. 3. Ecosystem management–Political aspects. 4. Ecosystem management–Monitoring. 5. Wildlife monitoring.
I. Title.
 QH75.H29 2011
 333.95'16–dc22

                                                                                      2010040965

A catalogue record for this book is available from the British Library.

Print ISBN: 978-0-470-66113-0
ePDF ISBN: 978-0-470-97934-1
oBook ISBN: 978-0-470-97933-4
ePub ISBN: 978-0-470-97955-6

Set in 10/12pt Times by Thomson Digital, Noida, India
Printed in Singapore by Markono Print Media Pte Ltd

Dedicated to the next generation of ecosystem managers

CONTENTS

Part III   ASSESSMENT   201

13 Current capabilities and limitations of the politically realistic EMT   203

APPENDICES   209

Preface

This is a how-to book for finding the most politically acceptable but effective plan for managing an at-risk ecosystem. In this book, finding such a plan is accomplished by first fitting mechanistic political and ecological models to a data set composed of both observations on political actions that impact an ecosystem and observations on variables that describe the ecological processes that are occurring within it. Then, the parameters of these fitted models are perturbed just enough so as to produce desired ecosystem state endpoints. This perturbed model gives the ecosystem management plan needed to reach the desired ecosystem state. To construct such a set of interacting models, topics from political science, ecology, probability, and statistics are developed. These group decision-making models capture group belief systems in their structure and parameter values. Hence, perturbing parameters to achieve needed shifts in behavior to cause desired ecosystem responses is equivalent to asking the question: 'What is the smallest change in group belief systems that would cause behavioral changes towards the ecosystem that would, in turn, result in the ecosystem responding in a desired way?' By focusing on belief system change, the tool is ideally suited to a non command and control ecosystem management system. Such non hierarchical management systems describe many at-risk ecosystems including those that straddle country boundaries. The book's running example is of one such trans-boundary ecosystem management case: conservation of cheetahs across Kenya, Tanzania, and Uganda (East Africa).

To demonstrate the proposed management tool's wide applicability, a sketch of how the tool could be used to manage the world's remaining population of blue whales is given in Chapter 2. These two examples of managing at-risk species are appropriate for a book devoted to managing natural resources when biodiversity is viewed as a natural resource.

Two types of readers will get the most from this book. The first type of reader is a person who is in, or is training for, a job in environmental and/or wildlife management wherein one of the decreed management goals is the protection of some part of the ecosystem, for example, wildlife that is at threat from anthropogenic forces. This first type of reader might be a student in a natural resources management program – or a member of a forestry, fish and game, national park, environmental protection agency, or other conservation-focused agency. This reader might also be employed by a wildlife advocacy organization such as the African Wildlife Foundation or the World Wildlife Fund. The prerequisites needed by this first type of reader are some familiarity with natural resources and elementary statistics. This type of reader should read Chapters 1–5 to acquire a working knowledge of how to use the book's methods to manage an at-risk ecosystem. Section 4.3 in Chapter 4 does, however, contain material that is best understood by a reader possessing a knowledge of calculus-based probability and statistics along with the notion of a vector of random variables.

The second type of reader is one who is trained in both the social sciences and mathematical statistics and is interested in how social science theory, ecology, probability, inferential statistics, and computers can be synthesized to create a decision support system for the scientific management of an at-risk ecosystem. This second type of reader would typically be a student or academic in political science, political economy, ecology, natural resources management, or statistics. This type of reader would be best prepared by having some background in one or more of the areas of political science, ecology, or mathematical statistics. This reader should read all of the book's chapters in order.

This book has the following pedagogical features:

1. The East African cheetah management application of the proposed ecosystem management tool is used as a running example through all of the chapters.

2. Exercises are at the end of most chapters – making the book suitable for a graduate lecture course on natural resource and/or wildlife management.

3. A companion website (www4.uwm.edu/people/haas/cheetah_emt) contains all computer code and data used in this book. Specifically, this website contains the software for, and an example of, the book's main contribution: a web-based *Ecosystem Management Tool* (EMT). The following can be freely downloaded:

   - All software described in the book (namely, the **id** software package) in the form of Java source (.java) and Windows class (.class) files.

   - A user's manual for **id**.

   - The political actions data set for the East African cheetah EMT along with the data collection protocol and a suite of web-based data acquisition aids.

   - The ecosystem data set for the East African cheetah EMT.

   - Output files from (a) the East African cheetah EMT's ecosystem management plan search, (b) statistical estimation of the EMT simulator, and (c) the simulator's sensitivity analysis.

   - A web-based tutorial that covers the basics of probability, statistics, and influence diagrams.

   - Answers to all of the book's exercises.

List of Figures

List of Tables

Nomenclature

AER	Actual Error Rate
AERS	Actual Error Rate Sum
ASCII	American Standard Code for Information Interchange
BCOW	Behavioral Correlates of War
CITES	Convention on International Trade in Endangered Species
CPT	Conditional Probability Table
DAN	Document Archive Number
DBO	Desires, Beliefs, and Opportunities
DM-group	Decision-Making Group
DSA	Deterministic Sensitivity Analysis
ECA	Empirically Calibrated Agent-based model
EER	Expected Error Rate
EMAT	Ecosystem Management Actions Taxonomy
EMT	Ecosystem Management Tool
EOS	Earth Observation System
EPA	Environmental Protection Agency
FAO	Food and Agriculture Organization
GBIF	Global Biodiversity Information Facility
GCM	Global Climate Model
GDP	Gross Domestic Product
GIS	Geographic Information System
GUI	Graphical User Interface
GW	Global Workspace
ID	Influence Diagram
IntIDs	Interacting Influence Diagrams
IQR	Interquartile Range
IUCN	International Union for Conservation of Nature
IWC	International Whaling Commission
MC	Monte Carlo
MCMC	Markov Chain Monte Carlo

MDAS	Multiple Dimensions Ahead Search
MIT	Massachusetts Institute of Technology
MPEMP	Most Practical Ecosystem Management Plan
NGO	Nongovernmental Organization
OGA	Overall Goal Attainment
OSAPAER	One-Step-Ahead Predicted Actions Error Rate
OSARMSPE	One-Step-Ahead Root Mean Squared Prediction Error
PBSS	Parallel Best Step Search
PDPF	Probability Density Probability Function
POMP	Privately Optimal Management Problem
PRA	Probabilistic Reduction Approach
PSOSV	Parallel Search Over Subsets of Variables
RMSPE	Root Mean Squared Prediction Error
SA	Simulated Annealing
SDE	stochastic differential equation
SEM	Structural Equation Model
SML	Simulated Maximum Likelihood
SOMP	Socially Optimal Management Problem
subID	Sub-influence Diagram
TEEB	The Economics of Ecosystems and Biodiversity (Project)
UN	United Nations
USA	United States of America
USDA	US Department of Agriculture

Part I

MANAGING A POLITICAL–ECOLOGICAL SYSTEM

1

Introduction

1.1 The problem to be addressed

In this book, biodiversity is considered a nonrenewable natural resource (USAID 2005, p. 6, USGS 1997, Wikipedia 2010, UFZ 2008). Many species are headed for extinction in habitats that straddle two or more developing countries. With our current understanding of biological processes (circa 2010), the loss of a species is irreversible. Because of this irreversibility, it can be argued that this problem should be of high priority to all countries. This book gives one way to address this problem.

Two characteristics of this problem make solutions difficult to find. First, within developed countries, constituencies prefer their policy makers to spend most of their conservation budget on internal conservation programs. Because of this internal focus, developing countries, with inadequate budgets for conservation programs, can expect to receive (currently) only modest supplemental conservation resources from developed countries. Second, because the habitat of many at-risk species straddles the political boundaries of several developing countries, conventional wildlife conservation strategies (such as government-run command and control programs) may not be implemented with sufficient completeness to achieve a species' long-term survival.

These considerations have motivated the development here of an approach to ecosystem management that does not assume central control but instead, after building scientific models of both the political processes at work in the habitat-hosting countries and the dynamics of the ecosystem in which the managed species is a participant, searches for politically feasible management plans. In other words, this book proposes a two-step procedure: first understand how the political–ecological system works at a mechanistic level and only then begin a search for management

Improving Natural Resource Management: Ecological and Political Models   Timothy C. Haas
© 2011 John Wiley & Sons, Ltd

plans that require the least change in human belief systems in order to effect behavioral changes that result in a sequence of actions that leads to the survival of the species being managed. The term *political–ecological system* is used rather than *socio-ecological system* to emphasize the active, institution- and ecosystem-changing tendencies of human groups across an ecosystem.

Such an approach to ecosystem management is different from so-called 'adaptive management' because it emphasizes a positivist and reductionist understanding of the entire political–ecological system before attempts are made to manipulate it. Adaptive management, on the other hand, can be viewed as a sequence of ecosystem management experiments that are conducted with the hope that a successful strategy will be found before the managed species becomes globally extinct. For example, Moir and Block (2001) argue that adaptive management's eight-step cycle of Propose Actions, Form Hypotheses, Determine Data Needs, Design Monitoring Program, Install Monitoring Program, Monitor, Analyze Collected Data, Implement a Management Action results in a monitoring protocol over a time scale that is not derived from an understanding of the ecosystem's dynamics but, rather, is short in duration so that feedback (adaptation) can be used to possibly adjust the management plan. These authors argue that this forced short time interval in the feedback loop invites 'False Effects' to drive management action revisions. Further, many applications of adaptive management depend on statistical hypothesis testing which, in turn, usually relies on linear statistical models of ecosystem processes rather than mechanistic, (possibly) nonlinear models of ecosystem dynamics that may be dominated by cycles with long periods (Moir and Block 2001).

But in this author's view, wildlife management is in a state of crisis. Environmental degradation and loss of biodiversity are occurring at unprecedented rates while efforts to stem this often-irreversible damage are on the whole inadequate. Funding for these problems, however, is low relative to other fields such as defense or human health. In developing countries, where most species reside, such funding is glaringly inadequate. The ecosystem management problem on the other hand is complex in that effective management strategies need to take into account how political realities impede or promote the implementation of options that could protect an ecosystem. If a freely available system, based on the best available science, existed and was capable of finding politically feasible but effective (this book's definition of 'practical') ecosystem management plans, managers and observers of at-risk ecosystems could use this tool to develop specific, defensible proposals for stemming this destruction. Because of their practicality, these proposals would have the best chance of actually being implemented.

There is a dearth of books that combine the social sciences and conservation and few individuals have training in both areas. The need to integrate the social sciences and conservation disciplines, however, has been recognized by the conservation community, see Fox *et al.* (2006) and Liu *et al.* (2007) for extended discussions of this deficiency.

Decisions to actually implement an ecosystem management policy typically have a political component. The majority of current ecosystem management research, however, is concerned with ecological and/or physical processes.

Management plans that are suggested by examining the output of these models and/ or data analyses may not be supported by the affected human population unless the option addresses the goals of each involved social group (hereafter, *group*).

As a step towards meeting this need, this book describes an Ecosystem Management Tool (EMT) that links political processes and political goals to ecosystem processes and ecosystem health goals. Because of this effort to incorporate the effects of politics on ecosystem management decision making, the EMT described in this book is referred to as a *politically realistic EMT* – or simply the *EMT*. This tool can help managers identify ecosystem management plans that have a realistic chance of being accepted by all involved groups and that are the most beneficial to the ecosystem. Haas (2001) gives one way of defining the main components, workings, and delivery of an EMT (referred to there as an Ecosystem Management System). The central component of this EMT is a quantitative, stochastic and causal model of the ecosystem being managed and the social groups involved with this management. This model is called the *political–ecological system simulator* (hereafter, *simulator*). In this simulator, group decision-making models and the ecosystem model are developed in a probabilistic form known as an *influence diagram* (ID) (see Pearl 1988, p. 125). The other components of the EMT are links to data streams, freely available software for performing all ecosystem management computations and displays, and, lastly, a web-based archive and delivery system for the first three of these components.

The two main uses of the EMT are first to find practical ecosystem management plans, and second to allow any literate person with access to the Web the ability to assess for themselves the status of a species being managed with the EMT. This second use is intended to make more accessible to developed countries the status and challenges of managing critical ecosystems in distant, developing countries.

A core message of this book is that ecosystem analyses and optimal management plan studies cannot be one-off and performed at only one time point. Rather, such applied ecosystem research needs to be on going and constantly updated. Present journal publishing practices encourage one-off studies but ecosystems are on going and dynamic. The tools contained in this book's EMT are in part meant to make such on going analysis easier to perform repeatedly and more cost effective in both hardware and labor.

1.2   The book's running example: East African cheetah

To fix ideas and to show feasibility, a politically realistic EMT for the management of the cheetah (*Acinonyx jubatus*) in a portion of East Africa is developed and applied as a running example throughout the book. The cheetah is listed as *vulnerable* in the Red List of Threatened Species maintained by the International Union for Conservation of Nature (IUCN) (Cat Specialist Group 2007). The portion of East Africa studied in this example is the land enclosed by the political boundaries

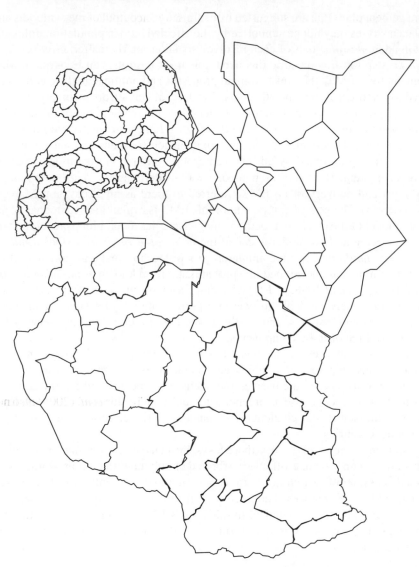

Figure 1.1   Area of East Africa that is the subject of the politically realistic East African cheetah EMT.

of Kenya, Tanzania, and Uganda (Figure 1.1). This ecosystem involves at least the cheetahs themselves, their prey, and the habitat in which these animals live. Humans are also a part of this ecosystem but here are modeled separately from the nonhuman aspects of the ecosystem. Specifically, along with an ID of the ecosystem, this EMT's simulator represents the following groups: (a) within each of the countries of Kenya, Tanzania, and Uganda, the president's office, the agency charged with executing wildlife and/or habitat protection actions (referred to herein as the *environmental protection agency* (EPA)), nonpastoralist rural residents (hereafter, *rural residents*), and pastoralists; and (b) a group of nongovernmental organizations (NGOs) that seeks to protect biodiversity within these countries (hereafter, *conservation NGOs*). This example was chosen in part to demonstrate the feasibility of applying the EMT to an at-risk species whose habitat ranges across several developing countries.

1.2.1 Background

Cheetah preservation is a prominent example of the difficulties surrounding the preservation of a large land mammal whose range extends over several countries. The main threats to cheetah preservation are loss of habitat, cub predation by other carnivores, and being shot to control predation on livestock (Gros 1998, Kelly and Durant 2000).

Kelly and Durant (2000) note that juvenile survival is reduced by lion predation inside wildlife reserves because these reserves are not big enough for cheetahs to find areas uninhabited by lions. Over crowding of reserves in Africa is widespread (see O'Connell-Rodwell *et al.* 2000) and cheetahs do not compete well for space with other carnivores (Kelly and Durant 2000). Although many cheetahs are currently existing on commercial land, this coexistence with human economic activities may not be a secure long-term solution for the cheetah. Bashir *et al.* (2004) also note that cheetahs do not compete well with lions and hyenas in protected areas (reserves or national parks) – and hence their survival in open areas and farmlands is crucial to their overall survival. These authors note, however, that cheetahs outside protected areas run the risk of being shot or poisoned by (a) trophy hunters for their skins or (b) farmers and pastoralists because they occasionally prey on their goats and calves.

One albeit expensive solution would be larger reserves that are free of poachers – possibly enclosed by an electrified fence. Such a solution was found to be the most viable for keeping elephants from destroying crops in Namibia (see O'Connell-Rodwell *et al.* 2000). Pelkey *et al.* (2000) also conclude that reserves with regular anti-poaching and anti-logging patrols are the most effective strategy for African wildlife and forest conservation.

A large portion of cheetah range is controlled by Kenya, Tanzania, and Uganda (see Kingdon 1977). In this range, cheetahs prey mostly on herbivores. Kingdon notes that because cheetahs take their prey via a strangulation bite attack, they have little success with prey that weigh more than about 60 kg. For this reason, cheetahs typically prey on the impala *Aepyceros melampus* (40 kg), Thomson's gazelle *Eudorcas thomsonii* (15 kg), Grant's gazelle *Nanger granti* (40 kg), lesser kudu *Tragelaphus imberbis* (40 kg), and gerenuk *Litocranius walleri* (25 kg) (Kingdon 1977). The average mass of these cheetah-prey herbivores is 32 kg. Hereafter, herbivores that weigh less than 60 kg will be referred to as *prey*.

Currently, the poverty rates in Kenya, Tanzania, and Uganda are 52%, 35.7%, and 44%, respectively. The adult literacy rates are 90%/79% (males/females) for Kenya, 85%/69% for Tanzania, and 79%/59% for Uganda (World Resources Institute 2005). With close to half of the population living in poverty, many rural Africans in these countries feel that conservation programs put wildlife ahead of their welfare and that large mammals are a threat to their small irrigated patches of ground and their livestock (Gibson 1999, p. 123). For these reasons, many such individuals are not interested in biodiversity or wildlife conservation.

Gibson (1999, p. 122) finds that the three reasons for poaching are the need for meat, the need for cash from selling animal 'trophies,' and the protection of livestock. Gibson's analysis suggests that to reduce poaching, policy packages need to be instituted that (a) deliver meat to specific families, not just to the tribal chief, (b) increase the enforcement of laws against the taking of trophies, and (c) improve livestock protection.

1.3   The EMT simulator

The simulator functions by having each group implement an action chosen from a predetermined repertoire that maximizes a multiple-goal utility function (specifically, the weighted sum of goal utility functions in which weights reflect relative goal importance). A temporal sequence of actions taken by those groups that affect the ecosystem (the result of playing this sequential game) is called an *ecosystem management plan*. Such an *actions history* may or may not be the result of a formal, articulated policy for managing the ecosystem.

1.3.1   Characteristics of an ideal simulator

To be convincing to all stakeholders, the EMT simulator needs to have the following two characteristics:

1. *Usability*: because of its predictive and construct validity, the simulator contributes to the ecosystem management debate by delivering insight into how groups reach ecosystem management decisions, what strategies are effective in influencing these decisions, how ecosystems respond to management actions, and which management actions contribute to ecosystem health. In other words, by running different management scenarios through the model, stakeholders both within and outside the ecosystem-enclosing countries are able to learn how political beliefs and actions need to change to improve measures of ecosystem health such as achieving the preservation of a threatened species.

2. *Clarity*: the simulator's construction and operation can be understood by individuals having a wide range of educational backgrounds.

These two simulator characteristics are seen as the most important for the development of a useful ecosystem management decision support system and are in agreement with recommendations given in Miles (2000).

For descriptions of predictive and construct validity see Feinsten and Cannon (2001), Babbie (1992), or Carmines and Zeller (1979). A model that possesses predictive validity displays prediction error rates that are lower than that of blind guessing. Here, predictive validity will be assessed with the simulator's one-step-ahead prediction error rate wherein, at every step, the simulator is refitted with all available data up to but not including that time step.

A model that possesses construct validity uses relationships, functions, and mechanisms that operationalize the current state of understanding of how groups reach decisions and how ecosystems unfold through time (ecosystem dynamics). Here, construct validity will be assessed by the degree to which the simulator's internal structure (variables and inter variable relationships) agrees with current theories of group decision making and mathematical models of wildlife population dynamics.

There is a tension between predictive and construct validity in that the development of a model sufficiently rich in structure to represent current theories of group decision making and ecosystem dynamics can easily become overparameterized, which, in turn, may reduce its predictive performance. The approach taken here is to develop a simple model that is faithful to theories of how groups reach decisions and to theories of ecosystem dynamics – followed by a fit of this model to data to help maximize its predictive performance.

At present, theories of group decision making and ecosystem dynamics are evolving. Models, then, will need to be modified and re-evaluated from time to time to incorporate advances in our understanding of how these processes work. A method is needed for determining whether a proposed model modification that improves the model's construct validity is also consistent with observations. For this purpose, a Monte Carlo (MC) hypothesis-testing procedure has been built into this book's EMT that allows an analyst to use statistical hypothesis testing to assess such modifications.

1.4   How to use the EMT to manage an ecosystem

The procedure for developing and using the EMT to manage an ecosystem is as follows:

1. Construct a stochastic model of each group's decision-making activity.

2. Construct a stochastic model of those elements of the ecosystem that are to be managed.

3. Collect data on group actions and on the output nodes of the ecosystem model.

4. Use this data to estimate the values of all parameters in these models.

5. Decide on ecosystem state goals.

6. Compute the *Most Practical Ecosystem Management Plan* (MPEMP) for these goals.

7. Enact the command elements of the MPEMP and execute activities that are intended to cause belief structure change towards the needed parameter values given in the MPEMP.

8. Continue to collect data and recompute the MPEMP as new data is acquired.

1.4.1   Ecosystem state goals

Step 5 involves the specification of desired ecosystem states in the future. The ecosystem state studied in the running example is the long-term survival of a species. Such a goal needs to be expressed stochastically since the simulator's ecosystem model is stochastic. Therefore, this goal is expressed herein as 'A species has a low risk of extinction in the future.' The definition of *low extinction risk* is given below.

1.4.1.1   One definition of *low extinction risk*

Although genetic variation concerns are important, for example Frankham *et al.* (2002), for purposes of easy interpretation, the phrase *low extinction risk* will mean herein that the probability of a species population falling below 10 animals 50 animal generations into the future is less than .01. Use of a number-of-generations definition of time accommodates differences of species lifespan in the assessment of extinction risk (Armbruster *et al.* 1999). The average lifespan of a cheetah in the wild is about 6.9 years (Honolulu Zoo 2008). Hence, cheetah abundance predictions with attendant measures of uncertainty would need to be computed about 350 years into the future.

1.4.2   No valuation of ecosystem services

No attempt will be made in this book to assign a value to natural resources such as biodiversity. Ecosystem state goals are identified exogenously to the proposed EMT. Of course, having the goal of preserving a species implies a value judgment. But the proposed EMT does not need estimates of the value of a species before it can be used to find the MPEMP. Rather, it only needs to be given desired ecosystem endpoints.

There is a large body of knowledge on how to assign value to natural resources, for example, The Economics of Ecosystems and Biodiversity (TEEB) project (TEEB 2010). Apart from brief discussions of these ideas in Chapter 4, the present work will avoid such efforts. The reason for this downplaying of ecosystem valuation is that, as the cheetah example will illustrate, different groups place different values on the same natural resource. Pastoralists in East Africa see live cheetahs as a liability to their livestock (negative value). Poachers in East Africa see value in a harvested cheetah and, by their actions, no value in future generations of cheetahs. Tourists and those who pay to watch wildlife television programs see value in a live cheetah. Whose valuation should be used? What markets exist along with legally enforced rights of ownership to make such valuations real in terms of hard currency? This author will make no attempt to answer these questions.

1.5   Chapter topics and order

Chapter 2 contains a sociological argument for the use of an interacting-groups-and-ecosystem approach to the simulator's construction. Then, computational details are given of how group IDs interact with each other and with the ecosystem ID over time. The book's running example of cheetah management in Kenya, Tanzania, and Uganda, referred to as the *East African cheetah EMT*, is introduced in this chapter.

Chapter 3 contains a short, self-contained example of an EMT for managing the global population of blue whales (*Baleanoptera musculus*). The intent of this chapter is to give the reader an overview of how an EMT is constructed from the identification of the at-risk species, development of models of involved groups, the ecosystem model, and data sources for group actions and ecosystem outputs.

A method for finding the MPEMP with a simulator that has been fitted to data is given in Chapter 4 along with an application of the method to the management of the East African cheetah. The MPEMP was first described in Haas (2008a). Although this method relies on first statistically fitting the simulator to data, it is presented before the statistical fitting chapter so that the primary use of a politically realistic EMT can be shown to the reader as early in the book as possible.

Chapter 5 contains a description of the book's web-based EMT and how it would be used to manage an ecosystem.

A review is given in Chapter 6 of some current theories of political decision making. Then, aspects of these theories are used to construct a general model of group decision making that is realized as an ID. Chapter 7 contains an application of the model developed in Chapter 6 to the presidential office, EPA, rural residents, and pastoralists within each of the countries of Kenya, Tanzania, and Uganda – and to a group of conservation NGOs operating in these countries.

A review is given in Chapter 8 of current differential equation models of wildlife population dynamics. Then, one of these models is used to construct an ecosystem ID that represents cheetah and prey population dynamics within the cheetah habitat that is conterminously enclosed by the political boundaries of Kenya, Tanzania, and Uganda.

The book's section on the statistical fitting of the simulator and its reliability assessment begins with Chapter 9. In this chapter, the protocol used to gather political data is given. The sources of ecosystem data used in the East African cheetah EMT appear in Chapter 10. This data consists of cheetah and prey abundance observations, vegetation type, and landuse – all by administrative district. The EMT's geographic information system (GIS) capabilities are used to display this data set. In Chapter 11, the model is statistically fitted to observations on group actions and wildlife abundance using an estimation method, called *consistency analysis*, that accounts for subject matter theory within the frequentist statistical estimation paradigm.

The simulator's parameter sensitivity is assessed and prediction error rates are computed in Chapter 12. This chapter also gives an MC hypothesis-testing procedure that can be used to improve the simulator's construct validity. Hypothesis testing can lead to erroneous conclusions when the data comes from an *observational study* (see Rosenbaum 2002) rather than a designed experiment. The size of this hazard can be ascertained by conducting a *sensitivity to hidden bias* analysis (also see Rosenbaum 2002) on the model and data set if a hypothesis test is computed to be significant. Chapter 12, therefore, also contains a review of this issue and one way to conduct a sensitivity to hidden bias analysis.

Current capabilities and limitations of the EMT are discussed in Chapter 13. This chapter also contains a plan for raising the training level of ecosystem managers. Managing natural resources is a complicated political–ecological problem and current graduate programs need to cover a wider range of material and at a higher level.

Exercises are included at the end of Chapters 2–4, 6–8, and 10–12 so that the book may be used in a lecture course on ecosystem management. Reading only Chapters 1 through 5 avoids most of the statistical material but should give the reader an outline of how the EMT is used to manage an ecosystem. Indeed, the intent of this organization is for the first five chapters to contain a concise overview of an EMT that is reinforced with examples – followed by a detailed user's manual for how to build an EMT, collect the needed data, and justify the political and ecological models that will form the EMT simulator.

1.6   The book's accompanying web resources

An extensive and free website supports the book's description of a web-based EMT. The site, www4.uwm.edu/people/haas/cheetah_emt, contains:

- All software described in the book (namely the **id** software package) in the form of Java source (.java) and Windows class (.class) files.

- A user's manual for **id**.

- The political actions data set for the East African cheetah EMT along with the data collection protocol and a suite of web-based data acquisition aids.

- The ecosystem data set for the East African cheetah EMT.

- Output files from (a) the East African cheetah EMT's ecosystem management plan search, (b) statistical estimation of the EMT simulator, and (c) the simulator's sensitivity analysis.

- A web-based tutorial that covers the basics of probability, statistics, and influence diagrams.

- Answers to all of the book's exercises.

2

Simulator architecture, operation, and example output

2.1 Introduction

To incorporate the interaction between groups and the ecosystem, an ID is constructed of each group (see Chapter 6) and one of the ecosystem (see Chapter 8). Then, decisions computed by each group ID at each time point that are optimal for that group at that time point are programmed to interact with the optimal decisions of other groups and with the solution history of the ecosystem ID. The model that emerges from the interactions of the group IDs and the ecosystem ID is referred to herein as an *interacting influence diagrams* (IntIDs) model. In this model, each group makes decisions that they perceive will further their own goals. Each of these groups, however, has a possibly inaccurate internal model of other groups and the ecosystem. In other words, an IntIDs model has groups implementing decisions to maximize their own utility functions by using (possibly) distorted internal representations of other groups and the ecosystem.

In this chapter, a description is first given of how an IntIDs model operates. Then, examples are shown of simulator output for the East African cheetah EMT. Upon review of Chapter 6, the reader should see how a group ID embedded in an IntIDs model constitutes an approach to the modeling of an agent that is similar to an empirically calibrated agent-based simulation model of social interactions advocated by Hedström (2005). Also, this chapter along with Chapter 11 should show the reader how an IntIDs model fitted with consistency analysis is similar to the

Improving Natural Resource Management: Ecological and Political Models   Timothy C. Haas
© 2011 John Wiley & Sons, Ltd

approach of Hedström except for the stronger statistical foundation of consistency analysis employed herein.

2.2   Theory for agent-based simulation

An IntIDs model is *actor oriented* and *mechanistic*. Such an architecture for modeling a sociological phenomenon is seen by Hedström (2005, Chapters 1–3) as the approach most likely to break the current logjam in the development of sociological theory. Hedström argues that sociology is primarily focused on how human behavior is affected by interactions with others through time. This focus on social interaction sets sociology off from psychology, which is more focused on individual cognition to explain behavior. Hedström reviews much of sociology research and finds a paucity of concrete, testable, theory of between-individuals interactions that can be realized in computational models to allow study of social interaction effects on behavior. Hedström predicts that until such models are developed and assessed against observation, sociology will not develop but remain an introverted discipline with empirical work being limited to the fitting of theoretically barren, statistical models borrowed from empirical psychology, such as regression or structural equation models (Hedström 2005, p. 151). Hedström concludes that the way forward is through the development of computer-based simulation models of between-individuals interactions through time that are fitted to observations (Hedström 2005, p. 149). He also gives a working example of one such approach: he and his colleagues have been developing a theory of how behavior is based on social interactions that he calls Desires, Beliefs, and Opportunities (DBO) theory (Hedström 2005, p. 9). He and his colleagues have implemented some of this theory in a computer simulation model and have calibrated some of the model's parameters to observations. He calls this calibrated model an empirically calibrated agent-based (ECA) model (Hedström 2005, p. 149).

Specific to the application presented in this chapter, Long and van der Ploeg (1994, pp. 64–65) argue for actor-oriented approaches to model the behavior of agrarian groups because:

One advantage of the actor approach is that one begins with an interest in explaining differential responses to similar structural circumstances, even if the conditions appear relatively homogeneous. Thus one assumes that the differential patterns that arise are in part the creation of the actors themselves. Social actors are not simply seen as disembodied social categories (based on class or some other classificatory criteria) or passive recipients of intervention, but active participants who process information and strategize in their dealings with various local actors as well as with outside institutions and personnel. The precise paths of change and their significance for those involved cannot be imposed from outside, nor can they be explained in terms of the working out of some inexorable structural logic, such as implied in de Janvry's (1981) model of the 'disarticulated periphery.' The different patterns of social organization that emerge result from the interactions, negotiations and social struggles that take place between the several kinds of actors. These latter include not only those present in given face-to-face encounters but also those who are absent but who nevertheless influence the situation, affecting actions and outcomes.

Jones (1999) applies a qualitative application of this approach to land degradation in Tanzania.

For purposes of modeling political systems in particular, Lewis (2007), in a review of the book *Complexity in World Politics* (Harrison 2006), mentions a failure of agent-based models of political systems to account for authority and other differential power relationships between agents since the existence thereof appears to conflict with agent autonomy. The group ID models of Chapter 6 resolve this issue through the use of the `Relative Power` node in each group ID.

2.2.1 Other agent-based social system simulators

Hedström's views notwithstanding (see above), there is debate about the potential of agent-based models to contribute to social science theory (Page 2008). The usefulness of such simulations, however, has been shown in the well-funded area of agent-based simulations of battlefield dynamics, for example, Li and Dang (2007), Yang *et al.* (2005), and US Army (2007). Although seemingly no more than a defense computer system, battlefield simulation is applied social science since it seeks to develop computational models of human cooperation within one army and human conflict between armies. Developers of these simulators, due to the urgency of their tasks, have not been able to wait for a final assessment of the value of agent-based modeling within the social sciences. Instead, these researchers have as their first priority the development of useful battlefield simulators and, as a second priority, advancement of the underlying social science theory of these simulation systems, for example, Saadi and Sahnoun (2006). The focus of the present book is similar: first, achieve a working and reliable politically realistic EMT for managing an ecosystem, and, second, provide methods that allow users to determine if some new social science theory might improve the reliability of the EMT simulator.

Anderies (2002) gives a basic theory of how artificial agents could be used to model interactions between organizations and ecosystems. Manson (2002) gives a general discussion of how a multiagent system could be validated. And a detailed model of an ecosystem is developed by Janssen *et al.* (2002).

2.3 Action messages and IntIDs model operation

2.3.1 Input–output nomenclature

Call a particular triad of input action, actor, and subject (of the input action) an *in-combination,* and a particular dyad of output action and target an *out-combination.*

An in-combination is a three-node combination of conditioning values, and the associated out-combination is an action–target optimal decision. Call an in-combination, out-combination pair an *in–out pair*. Specifically, a group ID has four input nodes **time**, **input action**, **actor**, and **subject** – but only the last three are used to form an in-combination. An out-combination consists of the values on the two output nodes: **output action** and **target**.

2.3.2   ID basics

An ID is a graphical representation of a multivariate probability distribution wherein nodes represent random variables and edges represent dependencies among these variables (see Pearl 1988, p. 125 and Nilsson and Lauritzen 2000). A subset of an ID's nodes that are interconnected is called a *subID*. Random variables indexed by root nodes (nodes having no parents) are assigned unconditional distributions; all other random variables are assigned conditional distributions; that are indexed by each unique set of values of the variable's parents. An ID graphically portrays dependencies, random components, and control points (here, nodes that represent actions and nodes that represent targets).

The symbols that make up such a graph have the following meanings: a *node* is any random or nonrandom variable in the model, a *circle* denotes a chance (random) node, a *double circle* denotes a deterministic node, a *square* denotes a decision node, and a *diamond* denotes a utility or value node.

2.3.3   Example of a group ID simulating an ecosystem management decision

Consider the decision problem that a rural resident faces: should he/she poach animals to supplement his/her family's protein intake and run the risk of being prosecuted for poaching? Figure 2.1 gives the ID that represents this decision problem. Tables 2.1 and 2.2 give the conditional probability tables (CPTs) for

Table 2.1 CPT for the `Goal: Feed Family` node.

Parent value	Acquire bushmeat	Don't acquire bushmeat
Set traps	.9	.1
Don't set traps	.001	.999

Table 2.2 CPT for the `Goal: Avoid Prosecution` node.

Parent value	Arrested for poaching	Not arrested
Set traps	.7	.3
Don't set traps	.001	.999

Table 2.3 Relative importance weights for the `Utility: Overall Goal Attainment` node.

FF value	AP value	
	Arrested for poaching	Not arrested
Acquire bushmeat	.8	1.0
Don't acquire bushmeat	.0	.5

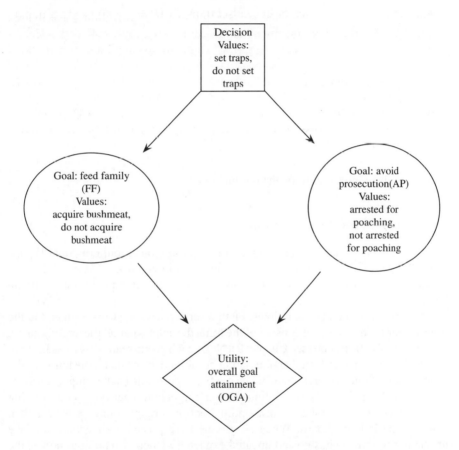

Figure 2.1   Illustrative example of a decision-making ID.

the two goal nodes, and Table 2.3 gives the utility node's importance weights. The entries in these tables are the parameter values for this ID. This set of values represents a decision maker who believes that there is a high chance of being caught but that the importance of acquiring meat is more important than being arrested for poaching.

Because the two goal nodes are independent, given a decision, the expected utility of setting traps is

$E[OGA|\text{set traps}] =$

.8$P(FF =$ acquire bushmeat|set traps)$P(AP =$ arrested for poaching|set traps)

$+ 0P(FF =$ don't acquire bushmeat|set traps)$P(AP =$ arrested for poaching
|set traps)

$+ 1P(FF =$ acquire bushmeat|set traps)$P(AP =$ not arrested|set traps)

$+ .5P(FF =$ don't acquire bushmeat|set traps)$P(AP =$ not arrested|set traps)

$=.8 \times .9 \times .7 + 0 \times .1 \times .7 + 1 \times .9 \times .3 + .5 \times .1 \times .3$

$=.789$

and for not setting traps,

$E[OGA|\text{don't set traps}]$

$= .8 \times .001 \times .001 + 0 \times .999 \times .001 + 1 \times .001 \times .999 + .5 \times .999 \times .999$

$= .5$

making the decision to set traps the optimal one.

2.3.4   IntIDs model operation

The political–ecological system simulator is based on a bulletin board (also referred to as a *blackboard* message posting architecture). Figure 2.2 gives such an architecture as it is used to construct the simulator in the East African cheetah EMT.

The simulator operates as follows. First, a seed *action message* is posted to the bulletin board. This message consists of the time, the actor's name, the target's name, and the EMAT action code (see Chapter 9). Next, each group reads this message and, after determining its preferred action–target combination, posts this reaction to the seed message. Time is incremented to the next time point and each group reads these messages and computes its own optimal action–target combination by conditioning on the values in the message. These optimal action–target combinations are then posted to the bulletin board. When all groups have posted their action message and the ecosystem ID has posted updated expected values of its output nodes, the time variable is incremented to the next time point and this process is repeated (see Figure 2.3).

This message posting algorithm allows feedback loops through time to emerge without the need for additional model structure. In addition, this modular structure makes it easy to build group IDs using different theoretical constructs and plug them into the simulator.

At each time point, each group ID computes a reaction out-combination to each recognized action message on the bulletin board that carries a time stamp equal to

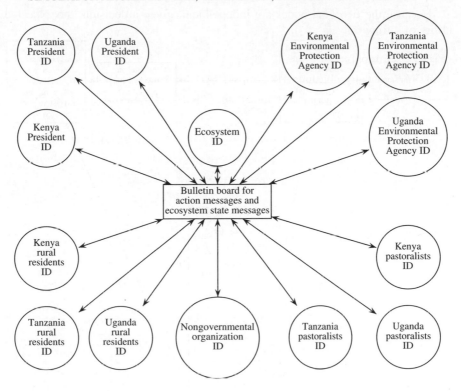

Figure 2.2   Schematic of the IntIDs model of interacting political and ecological processes.

the previous time point. Then, that group posts two action messages to the bulletin board. The first message is of an out-combination that gives that group the highest expected value on their OGA node. The second message is of an out-combination that gives the highest possible expected OGA value under the restriction that its target is different from that of the first posted message.

Groups that are directly affected by the ecosystem read an ecosystem state message at each time point that contains the expected values of the ecosystem variables at that time as computed by the ecosystem ID.

2.4   A plot for displaying an actions history

A new plot has been developed that displays a sequence of actions and reactions by several interacting groups. The form and interpretation of this plot is given here followed by an example.

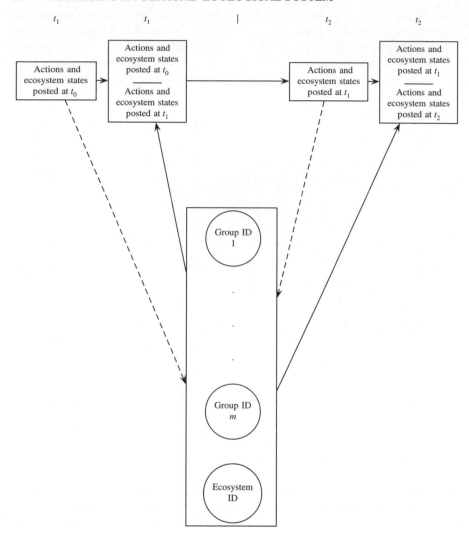

Figure 2.3   Sequential updating scheme of an IntIDs model consisting of m groups and an ecosystem. Bulletin board states are indicated by the boxes in the top row. These boxes contain action messages and ecosystem state messages that have been posted at the indicated time. A dashed arrow indicates messages are read but not removed. A solid arrow indicates message addition. Only two time points are depicted.

To develop a plot of either an observed or model-generated actions history, the concept of an *action–reaction pentad* is needed. First, let the group being studied be referred to as the *decision-making group* (DM-group). An action–reaction pentad has the form

input action–actor–DM-group–output action–target

where the DM-group is the target of the input action. An action–reaction pentad describes how a DM-group reacts to a particular in-combination. There is no passage of time between the input action and the output action.

If the input file is from an observed history of out-combinations, then the in-combinations are latent. Latent in-combinations can only be guessed at. One way to make such guesses is with an *apparent* in-combination. Apparent in-combinations are found by simply looking back from an action to the most recent action directed against that DM-group and *assuming* that it is the action to which the DM-group is reacting to. Note that only model output gives the **input action–actor** information associated (or causing) a DM-group's **output action–target** decision. Hence, use of this plot on an observed actions history is based on the above assumption.

2.4.1   Plot description

In this plot, an actions history is displayed by plotting action–reaction pentads on the vertical axis against the time value of the pentad's action on the horizontal axis. The plot symbol letter indicates the Ecosystem Management Actions Taxonomy (EMAT) category (see Chapter 9) of the action as follows: an 'm' indicates a militaristic action, a 'd' indicates a diplomatic action, an 'e' indicates an economic action, and an 's' indicates an action directed against the ecosystem.

Pentads at time t and $t + 2$ are connected by a solid line if the **action–target** of the time t pentad is the **input action–DM-group** of a pentad at time $t + 1$ and this time $t + 1$ pentad has an **output action–target** that is the **input action–DM-group** of some pentad at $t + 2$. This is called a *three-pentad sequence*. A horizontal line indicates that a target has reacted to an actor's action with a reaction directed back at this actor. Plotting pentads through time clearly displays *threads* of action–reaction sequences.

In a three-pentad sequence, the pentad at time $t + 1$ is not plotted but can be deduced by assembling the elements of the time t pentad and the time $t + 2$ pentad (see Table 2.4). Not plotting the middle pentad in a three-pentad sequence allows larger 'chunks' of the actions history to be plotted – thereby reducing the plot's complexity. Note, however, that three-pentad sequences only form when the subject of an in-combination is the DM-group itself.

Table 2.4 Reassembling the middle pentad in a three-pentad sequence.

	Time point	
t	$t + 1$	$t + 2$
Output action	Input action	
DM-group	Actor	
Target	DM-group	Actor
	Output action	Input action
	Target	DM-group

2.4.2 Vertical axis labels

An observed actions history typically has many more pentads than one generated by a model. Because of this potentially large number of pentad labels, the pentads of an observed actions history plot are sorted by DM-group and only the first pentad

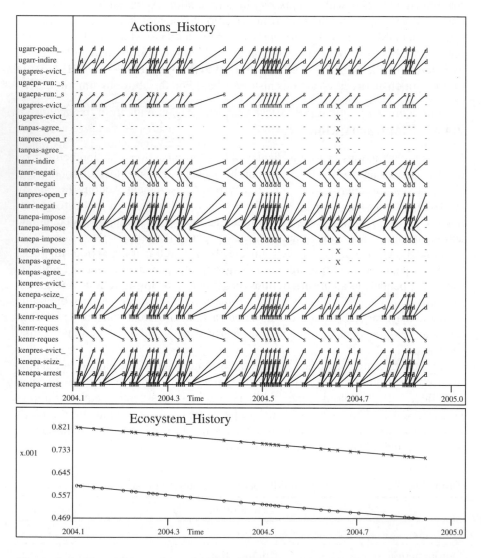

Figure 2.4   East African cheetah EMT simulator output over the year 2004. In the actions history plot, an 'X' symbol denotes a model-generated action that matches an observed one. Herbivore fraction detected and cheetah fraction detected are indicated by the 'o', and 'x' symbols, respectively. See the text for further details on plot symbol definitions and plot interpretation.

for each DM-group is labeled on the vertical axis. This label consists of only the DM-group's name.

For a model-generated actions history, every pentad is labeled by its DM-group—action pair. In addition, these pentads are ordered on the vertical axis so that total thread-length is minimized. Doing so makes it easier to see threads in the plot.

2.4.3   Example: output from the East African cheetah EMT simulator

An IntIDs model for the East African cheetah EMT is built from the group IDs given in Chapter 7 and the ecosystem ID of Chapter 8 using parameter values found by fitting the model to a political–ecological data set (see Chapter 11). This model is run over the period 1997 to 2009.

To show group interactions more clearly, Figure 2.4 gives the pattern of group-to-group interactions along with ecosystem output over just the year 2004 using the actions history plot described above.

Notice the reoccurring action of indirectly damaging wildlife habitat on the part of rural residents of all three countries. The IntIDs model exhibits a perfectly repeating pattern of interactions because the constituent group IDs currently do not have a memory mechanism and, due to the parameter values used in this example, are insensitive to changes in ecosystem state.

2.5   Conclusions

Agent-based simulation of social systems has been reviewed. Such an approach to modeling social systems may be an improvement over linear statistical models of aggregate measures of social systems – the dominant approach taken in social science modeling. Agent-based simulation forms the basis of a model of the political–ecological system. The approach involves models of homogeneous groups posting decisions to a bulletin board which are in turn read and reacted to by other groups and the ecosystem that is affected by these decisions. Use of this agent-based, message-passing model allows feedback loops and other nonlinear behavior to emerge rather than being imposed on the model by (say) additional mathematical forms.

Modeling is focused on smaller groups of people. These models attempt to capture the aspirations of these smaller groups without concern for how actions that stem from their efforts to reach these goals affect other groups or the ecosystem. In other words, an IntIDs model does not have any management knowledge such as monetary values for ecosystem services or willingness-to-pay values. This modeling approach is consistent with the directions described in Chapter 1:

a political–ecological system needs to be realistically modeled first before any use of such a model for ecosystem management is proposed.

An example of an IntIDs model has been introduced: the groups that impact the cheetah in East Africa. In order to show some output from this example IntIDs model, a new way was presented for plotting a history of political actions and reactions.

2.6   Exercises

1. Modify the parameter values in the example of simulated decision making of Section 2.3.3 so that not setting traps is the optimal decision.

2. Rerun the East African cheetah EMT simulator with different initial actions and compare your output to that of Figure 2.4.

3. Develop a schematic similar to Figure 2.2 that captures the interactions between groups that affect and are affected by the Mexican gray wolf in the states of Arizona and New Mexico in the USA.

3

Blue whale population management

3.1 Introduction

This chapter provides a self-contained example of how an EMT can be constructed for a political–ecological system that involves developed countries and a marine ecosystem. This example is intended to contrast with the book's running example of a terrestrial ecosystem managed by developing countries.

Background information, ecosystem-affecting groups, ecosystem models, and data sets are all described in this one chapter so that the reader can see a complete example of how this book's EMT is applied to a specific ecosystem management problem before delving into detailed presentations of each such aspect in Chapters 4 through 12 for the East African cheetah EMT.

3.1.1 Blue whale facts

The blue whale (*Balaenoptera musculus*) is a member of the Cetacea order, suborder Mysticeti (Stoett 1997, pp. 34, 45). An adult blue whale is between 70 and 120 feet (22 and 37 m) long and weighs anywhere from 70 to 120 tons (71 to 122 tonnes) (MacDonald 2001, Arkive 2009). This size and weight suggest that the blue whale may be the largest animal to have ever lived on earth – larger than the largest known dinosaur (Wikipedia 2009a). Blue whales use large elastic filter plates (baleen) to filter the water from a mouthful of krill before swallowing the krill whole as these whales have no teeth. Blue whales live for 30 to 90 years.

Improving Natural Resource Management: Ecological and Political Models   Timothy C. Haas
© 2011 John Wiley & Sons, Ltd

3.1.2   Some terminology

Whales are referred to as *cetacean creatures* and the set of controversies surrounding whale conservation is referred to as the *cetacean issue*. The hunting of whales offshore is referred to as *pelagic whaling*.

3.2   Current status of blue whales

3.2.1   Blue whale prevalence over the past 500 years

There were between 200 000 and 311 000 blue whales living in the Antarctic Ocean alone just as the factory ship harvesting system began in the 1930s. Before that, blue whales were not widely hunted for two reasons. First, a blue whale could usually swim faster than the rowed harpoon boats used to hunt them and, second, for those whales that were harpooned, the carcass was too big to be processed by a typical whaling ship of the time – making this species of whale unattractive for commercial harvesting. But in the 1920s and 1930s, faster boats with grenade-tipped harpoons were able to run down blue whales, kill them, and then, using a buoyancy device, tow them back to large factory ships. These factory ships, in turn, winched the carcasses up a stern-fitted ramp to internal processing facilities. This new system achieved the near annihilation of the species by harvesting about 378 000 blues whales from the Pacific, Antarctic, and Indian Oceans between 1930 and 1970 (Wikipedia 2009a).

The IUCN lists the blue whale as endangered on its red list (IUCN 2009), the Convention on International Trade in Endangered Species (CITES) lists the blue whale in its Appendix I: Endangered (CITES 2010), and the US Marine Fisheries Service lists the species as endangered (Arkive 2009, Wikipedia 2009a).

Unlike many other species, there is no captive population of blue whales that could be drawn upon to reintroduce the species should it go extinct in the wild.

3.2.2   Ecosystems to which blue whales belong

Today, approximately 1000 to 2500 blue whales exist in three main areas: the northeast waters of the Pacific Ocean, the Antarctic Ocean, and the Indian Ocean (Wikipedia 2009a) (see Figure 3.1).

Because of reversed seasons, the northern hemisphere blue whale population has little contact with the southern hemisphere population.

3.2.3   Current causes of blue whale mortality

A moratorium on the commercial harvesting of blue whales was passed by the International Whaling Commission (IWC) in 1965 (Stoett 1997, p. 157). So far, all

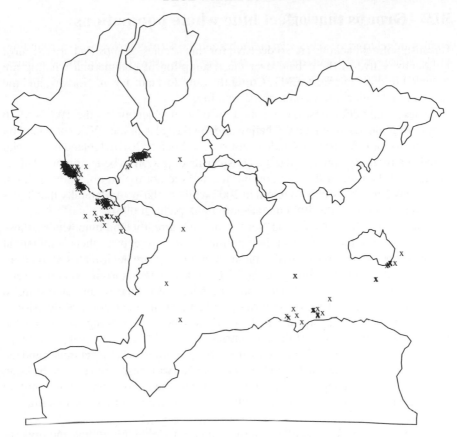

Figure 3.1   Blue whale sightings between the years 1965 and 2010. Data retrieved from the Global Biodiversity Information Facility (GBIF). See Section 3.6.2 for details of the database query. Continental boundaries were obtained by first converting the United Nations PDF map of the world (UN 2010) to a GIF file with Adobe Photoshop – and then applying **id**'s *digitizing system to this GIF file.*

countries have respected this moratorium. It is believed that only a small number of blue whales are taken illegally.

Blue whales are also struck by ships at sea. Most strikes are made by US Navy vessels. Such strikes are rare, however, because there are so few blue whales (IUCN 2009). If the blue whale population recovers, such strikes will have more of an effect on their abundance.

Because blue whale population sizes are so small, the biggest threat to their long-term survival is a premature resumption of commercial harvesting of this species. Political actions by either NGOs such as the IWC or individual countries that change the political and/or economic costs of such whaling will be the main determinants of the blue whale's fate.

3.3   Groups that affect blue whale populations

Because political actions are so decisive for this species, groups that are the most politically active in the debate over the resumption of commercial whaling are modeled in this illustrative EMT. Currently, only Iceland, Japan, Norway, and the Russian Federation allow commercial whaling.

These same countries, having filed an official objection to the IWC's 1986 moratorium on all commercial whaling, are, by the rules of the IWC, not bound by the moratorium (Australian Whale Conservation Society 2010). Iceland is a special case because many countries do not consider its objection to be legitimate. This is because, at the time of the moratorium, Iceland did not file an objection. Then, it left the IWC in 1992 – only to rejoin in 2002 while declaring unilaterally that it was filing an objection to the 1986 moratorium (Wikipedia 2009b).

Although Iceland, Japan, and Norway do not currently hunt blue whales, these countries are the most active in their political efforts to expand their commercial whaling operations and to make these operations more legitimate internationally – mainly through influencing the IWC to end its moratorium on commercial whaling. In other words, all that would be needed for blue whale harvesting to begin anew by any of these countries would be an internal political decision – perhaps in the form of a nonzero blue whale catch quota being passed by the legislature of one or more of these countries.

Hence, below, group IDs are constructed of the US anti-whaling complex and the pro-whaling complexes of Iceland, Japan, and Norway, respectively. The Russian Federation does not appear to be actively lobbying the international community to condone its return to commercial whaling – and hence is not part of this illustrative EMT.

Here, a 'complex' is a collection of cooperating lobbying groups and government agencies. The term was coined to describe the political cooperation between the US defense industry and the US Department of Defense: the so-called *US military–industrial complex* (Eisenhower 1961). An *advocacy coalition* within the field of environmental politics is a similar concept (see Sabatier and Jenkins-Smith 1993 and Weible 2007).

Model-definition files of each group described below are available for download at the blue whale EMT's website, `http://www4.uwm.edu/people/haas/blue_whale_emt/`.

3.3.1   Anti-whaling complex of the USA

This complex consists of anti-whaling lobbying groups such as Greenpeace USA (the US arm of Greenpeace International), the US Department of Commerce, and the US president. The US Department of Commerce enforces the Fisherman's Protective Act and the Packweed/Magnuson Amendment to the Fishery Conservation Management Act. The Protective Act bars fish product imports from countries certified by the Secretary of the Department of Commerce as having violated marine animal conservation agreements to which the USA is a signatory, for example, the IWC's moratorium on whaling. The Fishery Conservation Management Act blocks access into the 200 mile (320 km) band of US coastal waters by countries that violate marine animal conservation agreements to which the USA is a signatory (Stoett 1997, pp. 86–88). The president has the discretion of not enforcing the fishing ban against a certified country (Wikipedia 2009d). The Japanese, for example, take 75% of each year's Alaskan coastal bottom fish harvest (Stoett 1997, p. 87). The threat of these economic sanctions is the main reason why the IWC's whaling moratorium actually stops whales from being harvested.

3.3.1.1   ID

The group ID is as follows (see Figure 3.2). The sole audience is made up of anti-whaling activists in the USA. This complex has the two goals of Maintain Whaling Ban and Maintain Political Influence. This second goal is most affected by the perceived level satisfaction of the group of anti-whaling activists. The goal of maintaining the whaling ban is influenced by the Input Action. This is a purely political model in that there is no ecosystem status input node.

Actions by other groups that this group will react to are: (1) *lobby against anti-whaling IWC motions*, (2) *lobby for pro-whaling IWC motions*, (3) *vote against anti-whaling IWC motions*, (4) *vote for pro-whaling IWC motions*, (5) *raise this year's whale catch quota*, (6) *lower this year's whale catch quota*, (7) *catch more whales than the previous year*, and (8) *catch fewer whales than the previous year*.

Groups executing these input actions that are recognized by this group (called *actors*) are the pro-whaling complexes of Iceland, Japan, and Norway. Subjects of these input actions that are recognized by this group are the ecosystem and the pro-whaling complexes of Iceland, Japan, and Norway.

Actions that this group may take are: (1) *lobby for support of anti-whaling IWC motions*, (2) *lobby against pro-whaling IWC motions*, (3) *vote for anti-whaling IWC motions*, (4) *vote against pro-whaling IWC motions*, (5) *certify that a country has violated the IWC whaling moratorium*, and (6) *de-certify such a country*. Targets of these output actions are the anti-whaling complex of the USA (itself), and the pro-whaling complexes of Iceland, Japan, and Norway.

3.3.2   Pro-whaling complex of Iceland

Whaling in Iceland appears to have significant support from the Icelandic public (Wikipedia 2009b). The industry is economically focused on exporting whale meat to Japan as there is not sufficient local demand to support the industry. Japan, unlike many countries, places no restrictions on the importation of whale meat. Iceland's government and industry believe that their local stocks of minke (*Balaenoptera acutorostrata*), sei (*Balaenoptera borealis*), and fin (*Balaenoptera*

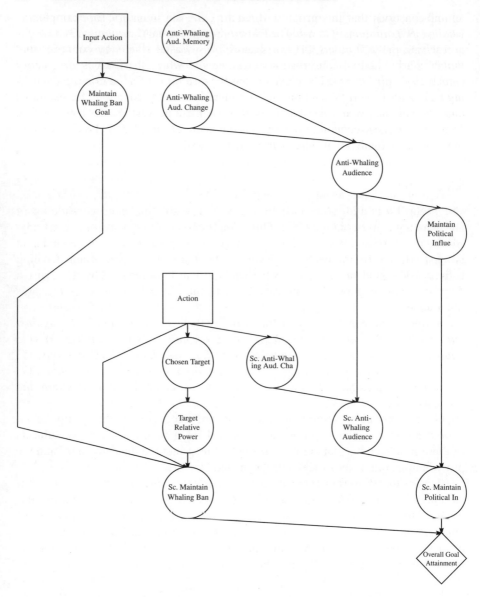

Figure 3.2   Anti-whaling complex of the US group ID.

physalus) whales can be harvested sustainably. Currently, Iceland does not harvest blue whales.

Iceland does not appear to be making a cultural issue of its whaling operations. Instead, it is focused on making money in the Japanese whale meat market. Hence, this group ID has no audiences and only one goal node: `Expand Commercial Whaling`.

Input actions that this group will react to are: (1) *lobby for support of anti-whaling IWC motions*, (2) *lobby against pro-whaling IWC motions*, (3) *vote for anti-whaling IWC motions*, (4) *vote against pro-whaling IWC motions*, (5) *certify that a country has violated the IWC whaling moratorium*, (6) *de-certify such a country*, (7) *raise this year's whale catch quota*, (8) *lower this year's whale catch quota*, (9) *catch more whales than the previous year*, and (10) *catch fewer whales than the previous year*. Actors executing these input actions are the anti-whaling complex of the USA and the pro-whaling complexes of Japan and Norway. Subjects of these input actions are the pro-whaling complex of Iceland (itself), the ecosystem, the anti-whaling complex of the USA, and the pro-whaling complexes of Japan and Norway.

Actions that this group may take are: (1) *lobby against anti-whaling IWC motions*, (2) *lobby for pro-whaling IWC motions*, (3) *vote against anti-whaling IWC motions*, (4) *vote for pro-whaling IWC motions*, (5) *raise this year's whale catch quota*, (6) *lower this year's whale catch quota*, (7) *catch more whales than the previous year*, and (8) *catch fewer whales than the previous year*. Targets of these output actions are itself, the anti-whaling complex of the USA, the pro-whaling complexes of Japan and Norway, and the ecosystem.

The graph of this group ID is similar to that of the pro-whaling complex of Japan, presented next.

3.3.3   Pro-whaling complex of Japan

The Japanese have respected the 1986 IWC moratorium on commercial whaling. At present, the Japanese restrict their whaling to a scientific program that is allowed under the IWC commercial whaling moratorium. No blue whales are taken under this program. The Institute of Cetacean Research carries out this whaling operation exclusively.

Politically, the Japanese government helps to fund the Institute of Cetacean Research and does not oppose the Japanese whaling industry's goal of the resumption of commercial whaling. The Japanese whaling industry's push for commercial whaling takes the form of voting against IWC moratorium measures, lobbying other countries to support the resumption of commercial whaling, informing the general public of its views through various media outlets, and maintaining an intact whaling industry through its IWC-allowed scientific whaling operation. Some of these lobbying efforts are carried out by the Japan Whaling Association, the mouthpiece for the Japanese whaling industry.

Whaling is an insignificant portion of Japan's economy. The Japanese public first started eating a lot of whale meat after World War II under the encouragement of US General MacArthur but lost its taste for the meat by the end of the 1960s (New York Times 2007). For example, a prominent whale meat restaurant closed due to disappointing sales (Demetriou 2008). Most Japanese have no strong opinions concerning whaling (British Broadcasting Corporation (BBC) News 2006). Why, then, is Japanese whaling an issue at all?

Some observers believe that the whaling industry, in order to survive, is attempting to connect the West's pressure on Japan to cease whaling activities with Western cultural imperialism wherein the West imposes environmental values on formerly conquered countries. The idea is that the West should not dictate to the Japanese that slaughtering pigs and cows is acceptable but not whales (BBC News 2006). According to these observers, the strategy of the Japanese whaling industry is to discourage the Japanese government from withdrawing support for the whaling industry's agenda of expansion since, if it does, it can be accused by the Japanese electorate of caving in to US-led cultural imperialism.

3.3.3.1   ID

One way to use an ID to represent this possible political situation is as follows (see Figure 3.3). First, the Japanese whaling industry and the Japanese government are aggregated into one group called the Japanese whaling industry–government complex. This group has the two goals of Expand Commercial Whaling Industry and Maintain Political Power.

Its one audience is the collection of vocal groups such as the Japan Whaling Association that are ostensibly concerned about insults to Japanese national pride by US-led cultural imperialism.

Input actions that this group will react to are: (1) *lobby for support of anti-whaling IWC motions*, (2) *lobby against pro-whaling IWC motions*, (3) *vote for anti-whaling IWC motions*, (4) *vote against pro-whaling IWC motions*, (5) *certify that a country has violated the IWC whaling moratorium*, (6) *de-certify such a country*, (7) *raise this year's whale catch quota*, (8) *lower this year's whale catch quota*, (9) *catch more whales than the previous year*, and (10) *catch fewer whales than the previous year*. Actors executing these input actions are the anti-whaling complex of the USA and the pro-whaling complexes of Iceland and Norway. Subjects of these input actions are the pro-whaling complex of Japan (itself), the ecosystem, the anti-whaling complex of the US, and the pro-whaling complexes of Iceland and Norway.

Actions that this group may take are: (1) *lobby against anti-whaling IWC motions*, (2) *lobby for pro-whaling IWC motions*, (3) *vote against anti-whaling IWC motions*, (4) *vote for pro-whaling IWC motions*, (5) *raise scientific whale catch limit*, (6) *lower scientific whale catch limit*, (7) *catch more whales than the previous year*, and (8) *catch fewer whales than the previous year*. Targets of these output actions are itself, the anti-whaling complex of the USA, the pro-whaling complexes of Iceland and Norway, and the ecosystem.

3.3.4   Pro-whaling complex of Norway

The Norwegian whaling industry is similar to that of Iceland. The domestic whale meat market is small and efforts to increase demand have been disappointing

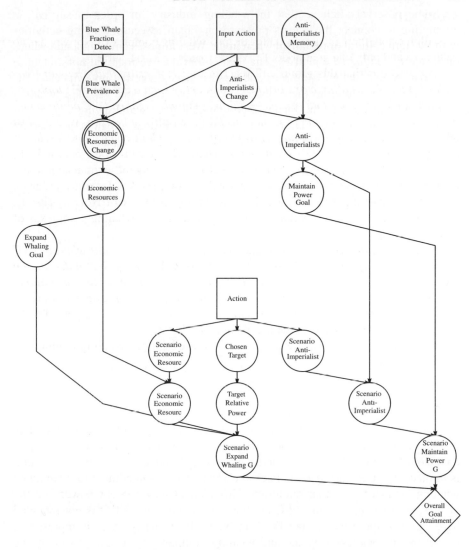

Figure 3.3   Pro-whaling complex of Japan group ID.

(eftec 2009). Norway appears committed to expanding its whaling industry through government subsidies (ScienceDaily 2009) – and through persistent lobbying of the international community to accept the idea that its government-set quotas on minke whales are sustainable (Wikipedia 2009c). Currently, Norway does not harvest blue whales. As with Iceland, Norway has exported some whale meat to Japan.

3.3.4.1   ID

As with Iceland, then, the group ID for the pro-whaling complex of Norway has no audiences and only one goal node: Expand Commercial Whaling.

Input actions that this group will react to are: (1) *certify that a country has violated the IWC whaling moratorium*, (2) *de-certify such a country*, (3) *lobby for support of IWC motions that are anti-whaling*, (4) *lobby against IWC motions that are pro-whaling*, (5) *vote for IWC motions that are anti-whaling*, (6) *vote against IWC motions that are pro-whaling*, (7) *raise this year's whale catch quota*, (8) *lower this year's whale catch quota*, (9) *catch more whales than the previous year*, and (10) *catch fewer whales than the previous year*. Actors executing these input actions are the US anti-whaling complex and the pro-whaling complexes of Iceland and Japan. Subjects of these input actions are the pro-whaling complex of Norway (itself), the ecosystem, the US anti-whaling complex, and the pro-whaling complexes of Iceland and Japan.

Actions that this complex may take are: (1) *lobby against anti-whaling IWC motions*, (2) *lobby for pro-whaling IWC motions*, (3) *vote against anti-whaling IWC motions*, (4) *vote for pro-whaling IWC motions*, (5) *raise whale catch quota*, (6) *lower whale catch quota*, (7) *catch more whales this year than the previous year*, and (8) *catch fewer whales this year than the previous year*. Targets of these output actions are the same as the list of subjects above.

The graph of this group ID is similar to that of the pro-whaling complex of Japan (see Figure 3.3).

3.3.5   IWC

As mentioned above, the IWC has no enforcement capabilities and hence any enforcement of its motions is through the use of trade sanctions imposed by member countries. Therefore, the IWC acts as a legitimizing NGO: as long as the moratorium is in place, the USA can use the Packweed/Magnuson Amendment to manipulate whaling behaviors of other countries. This lack of enforcement power and the heterogeneous nature of the IWC are the reasons why this NGO is not modeled here. Instead, the effect of the IWC on blue whale management is represented by (a) the actions taken by the anti-whaling complex of the USA to lobby the members of that NGO to support anti-whaling motions, (b) by the Japanese pro-whaling complex to lobby those same members to support pro-whaling motions, and (c) through the voting behavior of the USA, Iceland, Japan, and Norway on IWC motions.

3.4   Blue whale ecosystem ID

The model-definition files for the continuous-time blue whale population dynamics model described below are available for download at the blue whale EMT website.

3.4.1   Models of whale population dynamics

Clark (1976) studies the stability of an early finite difference model due to Allen (1963) called a *delayed recruitment* model. The IWC maintains a large finite difference model that stratifies across whale gender and age (Punt 1998). More recently, Mori and Butterworth (2004) give a finite difference model of blue and minke whale abundance. These authors speculate that if whale birth or death rates were made dependent on current abundance, the excessive oscillatory behavior of their model could be attenuated.

3.4.2   A continuous-time model

Currently, blue whale numbers are very low. Also, it is reasonable to assume that blue whales interfere very little with each other's gathering of prey. A continuous-time model is developed here that (a) is appropriate for the current dynamic of blue whale abundance, and (b) addresses the stability issue of the Mori and Butterworth (2004) model.

This model is described in terms of a general form for differential equation models of predator–prey population dynamics given by Yodzis (1994). This framework is as follows. Let B_t and N_t be the abundance of prey and predator at time t, respectively. Then

$$\frac{dB_t}{dt} = f(B_t) - N_t F(B_t, N_t) \tag{3.1}$$

$$\frac{dN_t}{dt} = N_t G(B_t, N_t) \tag{3.2}$$

where $f(B_t)$ is the prey's growth rate under no predation, $F(B_t, N_t)$ is the predator's *functional response*, and $G(B_t, N_t)$ is the predator's *numerical response*. To avoid unrealistic model behavior when predator abundance is low, Yodzis (1994) recommends $F(B_t, N_t)$ be defined as

$$\frac{t_h^{-1} B_t^n}{(Q t_h)^{-1}(N_0 + N_t)^m + B_t^n} \tag{3.3}$$

where Q, N_0, t_h, m, and n are positive-valued parameters. The parameter t_h is the amount of time a predator spends handling prey.

When predators do not interfere with each other's prey gathering (called *laissez-faire predator interaction*), a typical form for the numerical response function is $e F(B_t, N_t) - d$ where e and d are positive-valued parameters. The e parameter scales the predator's birth rate function, and the d parameter is a death rate that combines the effects of natural mortality and harvesting. One way to model the effect of time-varying whaling activity on blue whale abundance is to have d be time dependent.

Finally, a logistic growth rate for prey under no predation is reasonable for many real-world prey animals such as the krill prey of blue whales. In this case, $f(B_t) \equiv rB_t(1 - B_t/K)$ where r (prey birth rate) and K (prey carrying capacity) are parameters.

By adding the derivative of a Wiener process (in the sense of the Ito formula (Kloeden and Platen 1995, pp. 81, 92)) to Equations 3.1 and 3.2, the predator–prey model can be made stochastic. Doing so makes this model a *system of stochastic differential equations* (SDEs) (see Chapter 8).

Combining all of the above yields

$$\frac{dB_t}{dt} = rB_t(1 - B_t/K) - N_t \frac{t_h^{-1} B_t^n}{(Qt_h)^{-1}(N_0 + N_t)^m + B_t^n} + dW_t^{(B)} \qquad (3.4)$$

$$\frac{dN_t}{dt} = N_t \left[e \frac{t_h^{-1} B_t^n}{(Qt_h)^{-1}(N_0 + N_t)^m + B_t^n} - d_t \right] + dW_t^{(N)}. \qquad (3.5)$$

The modification suggested by Mori and Butterworth (2004) to reduce excessive oscillatory behavior of their model is accomplished by the N_t term in the denominator of the above form for the predator's functional response. This model could be extended by stratifying across predator gender and age.

A continuous-time model form is seen as crucial because (a) time is continuous, and (b) whale sightings and harvesting events happen at times that are typically both unpredictable and irregularly spaced.

3.4.3   ID

An ecosystem management action taken at time t is represented by one of the values of the *management* variable, M_t. Say that the effect of lifting the moratorium on commercial whaling is to be studied. Then, the simplest set of values for M_t would be *no harvesting (natural mortality only)*, and *harvesting at a rate thought to be sustainable*. The ecosystem is affected by M_t's influence on the d_t variable. The values of d_t are determined by a simple lookup function of M_t. Under M_t's first value, $d_t = .001$, and under its second value, $d_t = .1$.

The ecosystem ID, then, contains five nodes: t, M_t, d_t, B_t, and N_t as shown in Figure 3.4. In this figure, the undirected edge between prey and predator abundance depicts the dependence between these two nodes at time t. Koster (1997a, 1997b) shows that such a display of dependency between ID nodes is an exact representation of the probabilistic dependency between the associated random variables.

Given a particular time point t and particular management action taken at that time, the solution of the system of stochastic differential equations is a bivariate probability distribution for the random vector $(B_t, N_t)'$.

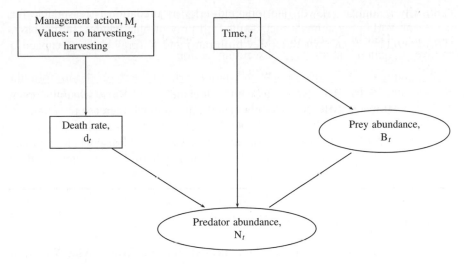

Figure 3.4   Ecosystem ID in the blue whale EMT simulator.

3.5   Interactions between IDs

Once the IntIDs model is constructed, the political–ecological system is simulated by computing a decision (**action–target** combination) for each group ID at a time point and solving the ecosystem ID up to that same time point. This procedure is repeated at each time point across a time interval. Time points are incremented by a time step of about two weeks. At each of these time points, **action–target** combinations are posted on a bulletin board. Then, each ID reads this **actor–input action–subject** triad and, if recognized, computes and posts its reaction.

Actions by group IDs at particular time points affect the ecosystem ID by changing the value of the natural mortality and harvesting parameter, d_t, in the blue whale population dynamics model at those time points. The ecosystem ID, after updating itself to reflect changes to this parameter value, affects group IDs through its updated value of blue whale prevalence. For example, the ID for the pro-whaling complex of Japan has a node that listens for a new value of blue whale prevalence. Such a group ID would subsequently make a decision that is, in part, influenced by this new blue whale prevalence value.

Groups also react to the actions of other groups. For example, if Japan were to engage in commercial whaling, the USA would certify Japan as having violated the IWC whaling moratorium.

3.6   Data sets for the blue whale EMT

The EMT website would maintain data sets used to statistically estimate the EMT simulator. These data sets would be collected from the sources described below.

Table 3.1 Example actions data for the blue whale EMT.

Document archive number	Date of action	Source of story	Group executing action	Target of action	Action
1	1 27 09	BBC News	Iceland	Iceland	Raise whaling quotes
2	9 25 09	BBC News	Iceland	Ecosystem	Increase fin whale catch
3	12 7 09	The Age	Norway	Norway	Raise whaling quotas
4	1 8 10	Associated Press	Japan	Sea Shepherd	Damage anti-whaling boat

3.6.1   Group actions data

Observations on group actions can be collected from the World Wide Web. For example, the protocol for collecting actions described in Chapter 9 is used to collect the actions given in Table 3.1 (see also www4.uwm.edu/people/haas/ blue_whale_emt).

3.6.2   Blue whale prevalence data

The blue whale sightings data of Figure 3.1 is one example of an open source of blue whale observations (GBIF 2010). This figure shows all blue whale sightings contained in the GBIF database from 1965 (the first year of the moratorium on commercial harvesting of the blue whale) to 2010.

3.6.2.1   GBIF citations

Figure 3.1 was constructed by accessing the following data sets through the GBIF data portal using http://data.gbif.org/datasets/resource/ followed by the data set's identification number. For example, the first data set is at http://data.gbif.org/datasets/resource/9111.

```
South Australian Museum Australia provider for OZCAM, (9111)
Museum Victoria provider for OZCAM, (9107)
SEAMAP - marine mammals, birds and turtles, (338)
NMNH Vertebrate Zoology Mammals Collections, (1837)
Seabirds of the Southern and South Indian Ocean, (70)
Whale catches in the Southern Ocean, (75)
Whale log - observations from ANARE voyages, (8210)
SOMBASE, (66)
Mammals of the Gothenburg Natural History Museum, (1047)
Mammals (NRM), (1041)
Colección Nacional de Mastozoología - Museo Argentino de Ciencias
Naturales 'Bernardino Rivadavia', (9115)
Cetacean distribution in the South Atlantic and South Pacific Ocean
(AR-OBIS) (OBIS South America, SOUTHERN OCEAN SUB-NODE), (417)
```

```
National Whale and Dolphin Sightings and Strandings Database, (79)
Western Australia Museum provider for OZCAM, (9113)
iziko South African Museum - Marine Mammal Collection (AfrOBIS), (403)
Australian Museum provider for OZCAM, (9105)
Paleobiology Database, (563)
Macaulay Library - Audio Data, (41)
Mammal Collection Catalog, (162)
Santa Barbara Museum of Natural History, (646)
Vertebrate specimens, (1795)
HMAP-History of Marine Animal Populations (CoML), (316)
MVZ Mammals Specimens, (8121)
PIROP (Shipboard Surveys), (171)
Canada Maritimes Regional Cetacean Sightings (OBIS Canada), (353)
Inventaire national du Patrimoine naturel (INPN), (2620)
Bay of Fundy Species List (OBIS Canada), (347)
Mammal Specimens, (124)
Take a Pride in Fife Environmental Information Centre - Records for
Fife from TAPIF EIC, (927)
Mammal collection, Natural History Museum, University of Oslo, (8067)
Taxonomic Information System for the Belgian coastal area, (361)
```

3.6.2.2   Other data sources

Another open source of blue whale observations is the OBIS-SEAMAP database maintained at Duke University (Read *et al.* 2010). This database contains observations up to about the year 2004.

3.7   Main points of this chapter's example

The intention of this chapter is to give an overview of how an EMT is constructed, what its main components look like, and the nature of data sets that would be used to fit its simulator. To this end, a self-contained example of an EMT is described herein that addresses the management of a species that has a global range and is threatened mostly by the actions of developed countries. This example shows one way to model the world-views of advocacy coalitions with IDs. A stochastic, continuous-time population dynamics model is also developed for the species to be managed, which in this example is the blue whale.

The next step in the development of this blue whale EMT would be to collect data on group actions and blue whale abundance so that the parameters of the EMT simulator could be statistically estimated. To support this task, open sources of political and ecological data streams are indicated above.

3.8   Exercises

1. Using **id**, build a group ID of an aboriginal group lobbying the IWC to allow it to engage in coastal, subsistence whaling. See Stoett (1997, pp. 117–121) for some background information.

2. Solve the deterministic form of the blue whale population dynamics model described above using Maple™ or MATLAB™. Use a constant value for the harvesting plus natural mortality parameter. Select reasonable parameter values that show the blue whale's population being nonzero for 50 blue whale generations. Then, find the smallest change in parameter values that results in the blue whale going extinct over this same time interval.

4

Finding the most practical ecosystem management plan

4.1 Introduction

Once the political–ecological system simulator has been fitted to a political–ecological data set, the simulator can be used to construct ecosystem management plans. In this book, a practical plan is one that demands the least change in group behavior patterns to achieve a desired level of ecosystem health as measured by the ecosystem ID's output nodes. This definition emphasizes political feasibility over a plan's expected profit or cost. Such a plan is referred to herein as the Most Practical Ecosystem Management Plan (MPEMP) and was introduced by Haas (2008a).

The layout of this chapter is as follows. First, some published methods for developing an ecosystem management plan are reviewed in order to clarify the differences between the MPEMP and extant notions of ecosystem plan optimality. As mentioned in Chapter 1, a statistical parameter estimation method called consistency analysis is used in Chapter 11 to fit the simulator to data. Because the MPEMP is defined in terms of this method, consistency analysis is briefly described before the MPEMP is defined and the procedure is given for its construction. The chapter concludes with a computation of the MPEMP for the East African cheetah EMT.

Improving Natural Resource Management: Ecological and Political Models   Timothy C. Haas
© 2011 John Wiley & Sons, Ltd

4.2   Some methods for developing ecosystem management plans

4.2.1   Leadbeater's possum

A recent example of a method for developing a command and control ecosystem management plan is the model due to Spring and Kennedy (2005) for the optimal management of a timber-producing forest that is also the habitat of an endangered possum. These authors arrive at a monetary value for the existence of a species (in this case, Leadbeater's possum, *Eymnobelideus leadbeater*) as follows. First, a mail survey is administered to a taxable population. Using the contingent valuation method, responses to questions in the survey are used to compute each respondent's willingness to pay for the long-term survival of the possum. Then, the possum is given an upper and lower bound 'existence value' by multiplying the associated extreme willingness-to-pay values across survey respondents by the size of the taxable population.

A number of timber stands are to be managed. Only old growth stands have habitats that are suitable for the possum. The actions of leaving alone or clearcutting each stand is simulated over many 50-year *stages*. Simultaneously, the possum's population dynamics is simulated across all stands. Timber values are computed at each stage along with the event of possum extinction in each stand. Under a discount rate, an exhaustive search is used to find a timber harvesting management plan (clearcut decision for each stand at each stage) that maximizes the sum of the present expected value of species existence and the present expected value of timber harvests, that is, the total present expected value of the ecosystem.

This approach to ecosystem management has three notable characteristics. First, the objective function to be optimized is the total present value of the ecosystem. Second, the model of ecosystem dynamics is assumed to be known and not subject to modification over the management time interval, that is, the model is not adapted in the light of experimental results as is allowed in adaptive management approaches. Third, species value is deterministically and positively linked to the sample statistics of a mail survey, the size of a taxable population, and the negative of the discount rate. Fourth, this scheme allows local extinction (a renewable resource).

4.2.2   Ecological/economic modeling

Brock and Xepapadeas (2002) consider the problem of managing an ecosystem that consists of several patches of habitat within which several species are competing for a limited resource. The only management activity allowed by these authors in their modeling is species harvesting. These authors solve two problems that involve humans: the privately optimal management problem (POMP) and the socially optimal management problem (SOMP). An inter-species resource competition model due to Tilman (1982) is integrated with an economic model of profits from species harvesting.

In the POMP, private agents control habitat patches and do not communicate with other agents or consider possible impacts of their actions on other patches when making harvesting decisions. For this problem, the authors find a solution that maximizes the present value of harvesting profits over an infinite time horizon. This solution is that only a single species is guaranteed to survive long term. This species survives because it is the one that generates the highest steady-state private profits.

In the SOMP, monetary values are placed on the abundance of species within patches by forming the product within a patch of a monetary value function of one unit of species biomass and the biomass of that species that is within the patch. The authors find a solution that maximizes the present value of the sum of harvesting profits and ecosystem state (called a *spectrum of ecosystem services*). This solution also guarantees only the long-term survival of a species that in the steady state returns the highest value of this sum.

Notable characteristics of this modeling approach are that (a) 'optimal' means largest monetary return, (b) the ecological model does not change over time, and (c) the value of the species can be expressed as a monetary value per unit of species biomass.

4.2.3   Adaptive management

As was first mentioned in Chapter 1, adaptive management is a command and control method for developing ecosystem management plans and, hence, is a third example of this top-down approach to ecosystem management. Adaptive management consists of two main variants: active and passive (Parma *et al.* 1998). Passive adaptive management keeps a family of models and uses a weighted average of model outputs as the output on which to make management decisions. Active adaptive management also maintains a family of models but adds at each stage a decision to experiment with the ecosystem in an effort to reduce model uncertainty. Both of these forms of adaptive management compute the present value of species harvesting profit with stochastic dynamic programming (Walters and Hilborn 1976).

Active adaptive management's side goal of reducing model uncertainty means that, under certain conditions, the method will harvest a species and risk local extinction in order to gain data on local extinction events so as to reduce the variance of the Bayesian posterior distribution of a parameter (Hauser and Possingham 2008). Such variance reduction is called *Bayesian learning* (see, e.g., Neal 1996, pp. 4–6). These harvest-to-learn decisions emerge as part of the solution to the active adaptive management optimization problem. Passive adaptive management, on the other hand, does not let parameter posterior distribution variance influence harvest decisions but does update the parameter's posterior distribution at each stage.

Adaptive management with parameter uncertainty, then, acknowledges model uncertainty and uses a sequence of Bayesian statistical estimates of the uncertain parameters to reduce this uncertainty as time goes by.

As with the nonadaptive command and control management methods above, 'optimal' in adaptive management means that the present value of harvesting profit is maximized. Hauser and Possingham (2008) find that when the time horizon is of medium length rather than infinite, precautionary management decisions (do not harvest the species) are optimal under active adaptive management. This finding is a formal way of saying that since the model is not known with sufficient accuracy, giving the species a buffer size by not harvesting it may be necessary to avoid extinction until model reliability can be improved and species size predictions can be made more accurately.

4.2.4   MPEMP compared to these methods

As described in detail below, MPEMP maximizes the agreement between group belief structures that are estimated from data and the belief structures that produce group actions that result in ecosystem states that are desired by the ecosystem manager. The optimality function is different: all of the above methods maximize ecosystem value while MPEMP seeks the least change in existing group belief structures. The above methods assume a renewable species population and hence allows local extinction to have positive probabilities. MPEMP, however, can be set up so that the chance of extinction is essentially zero. Both the above methods and MPEMP postulate some model of the ecosystem and some economic model of human decision making. Of the above methods, only adaptive management pays as much attention to model uncertainty as is paid in the construction of the MPEMP.

The most salient difference between MPEMP and the above methods, however, is that the above methods assume that the ecosystem manager has sufficient authority and resources to force the implementation of whatever sequence of management actions the optimal solution calls for. MPEMP, however, does not provide a sequence of commands to be executed to manage the ecosystem. Rather, MPEMP finds a new unfolding of future reality in which modified behaviors of several groups result in ecosystem responses desired by the ecosystem manager – even when some of these groups are outside the ecosystem manager's control. Hence, the MPEMP is an implicit, autonomous, grassroots, emergent, swarm-intelligent, decentralized ecosystem management plan.

4.3   Overview of the consistency analysis parameter estimator

IntIDs model parameters are estimated via a procedure referred to here as *consistency analysis*, described next. Let $\mathbf{U}^{(i)}$ be the vector that contains all of the chance nodes that make up the ith ID. IDs 1 through m are models of group perceptions and decision making while the $(m + 1)$th ID is the model of the ecosystem. Let $\boldsymbol{\beta}^{(Grp)} = (\boldsymbol{\beta}^{(1)'}, \ldots, \boldsymbol{\beta}^{(m)'})'$ be the stacked vector of group ID parameters wherein

$\beta^{(i)}$ parameterizes the ith group ID. Let $\beta^{(Eco)}$ parameterize the ecosystem ID. Finally, let $\beta = (\beta^{(Grp)'}, \beta^{(Eco)'})'$ parameterize the entire IntIDs model. Let $g_S^{(i)}(\beta^{(i)})$ be a *goodness-of-fit* statistic that measures the agreement of the ith ID's probability distribution (referred to here as the $\mathbf{U}^{(i)}|\beta^{(i)}$ *distribution*) and the sample (data set), S. Larger values of $g_S^{(i)}(\beta^{(i)})$ indicate better agreement. Specific forms of agreement functions are presented after this overview.

Each parameter in the model is assigned an a priori value derived from expert opinion, subject matter theory, or the results of a previous consistency analysis. Let $\beta_H^{(j)}$ be such a value assigned to the IntID's jth parameter. Collect all of these *hypothesis parameter values* into the *hypothesis parameter vector*, β_H. Let $g_H^{(i)}(\beta^{(i)})$ be the agreement between the distribution identified by the values of $\beta_H^{(i)}$ (the $\mathbf{U}^{(i)}|\beta_H^{(i)}$ distribution) and the $\mathbf{U}^{(i)}|\beta^{(i)}$ distribution. As with $g_S^{(i)}(\beta^{(i)})$, larger values of $g_H^{(i)}(\beta^{(i)})$ indicate better agreement. Note that $g_S^{(i)}(\beta^{(i)})$ is the agreement between a sample and a stochastic model – the $\mathbf{U}^{(i)}|\beta^{(i)}$ distribution – while $g_H^{(i)}(\beta^{(i)})$ is the agreement between two stochastic models – the $\mathbf{U}^{(i)}|\beta_H^{(i)}$ distribution and the $\mathbf{U}^{(i)}|\beta^{(i)}$ distribution.

Let $R_{g_S^{(i)}}^{(j)}$ be the range of the sample that is formed from all evaluations of $g_S^{(i)}$ from the beginning of the maximization procedure (see below) up to the jth evaluation of $g_S^{(i)}$. Define $R_{g_H^{(i)}}^{(j)}$ in a similar manner. The consistency analysis parameter estimator maximizes the function

$$g_{CA}^{(i)}(\beta^{(i)}) \equiv (1 - c_H)\left[\frac{g_S^{(i)}(\beta^{(i)})}{R_{g_S^{(i)}}^{(j)} + 1}\right] + c_H\left[\frac{g_H^{(i)}(\beta^{(i)})}{R_{g_H^{(i)}}^{(j)} + 1}\right] \qquad (4.1)$$

where $c_H \in (0, 1)$ is the analyst's priority of having the estimated distribution agree with the hypothesis distribution as opposed to agreeing with the empirical (data-derived) distribution. Haas (2001, Appendix) gives suggestions for assigning c_H. Let

$$\beta_C^{(i)} \equiv \arg\max_{\beta^{(i)}} \left\{g_{CA}^{(i)}(\beta^{(i)})\right\}$$

be the consistency analysis estimate of $\beta^{(i)}$. Hereafter, $\beta_C^{(i)}$ will be referred to as the *consistent* parameter vector for the ith ID.

Consistency analysis consists of the following four steps:

1. **Specify (the values for β_H).**
2. **Initialize (the values in β).**
3. **Maximize (the agreement function).**
4. **Analyze (the differences between hypothesis and consistent parameter values).**

The method's name comes from this final step: analyze the model's parameters by scrutinizing areas of the subject matter theory that had suggested hypothesis parameter values that are very different from their consistent values. Haas (2001, Appendix) outlines how this comparison of hypothesis and consistent parameter values should proceed.

One way to quickly locate conditional distributions for which hypothesis values are not in agreement with consistent values is to declare a disagreement between subject matter theory and observation if the mode of the associated node's conditional distribution specified with hypothesis values is different than when specified with consistent values.

The idea behind the definition of g_{CA} is to apply the c_H-based weights to *standardized* measures of agreement wherein the standardizing constants ($R^{(j)}_{g_S^{(i)}}$ and $R^{(j)}_{g_H^{(i)}}$) are each measure's range computed from all current computational experience. The range is selected to ensure that standardizing constants are always as large as possible during the course of the maximization procedure. Hence, statistical properties of the range statistic, such as its robustness, are not relevant in this application of the statistic since the agreement measures are not random variables, and values on these measures generated during the course of the maximization procedure are the results of a deterministic search.

One execution of consistency analysis produces a consistent parameter vector. Then, if a new data set is collected, statistical learning can be carried out with consistency analysis by setting the next hypothesis parameter vector equal to the current, consistent one before performing another consistency analysis on the new data set.

Consistency analysis can also be used to adjust an ID's utility weight coefficients that define its decision node. One example of using data to estimate a decision maker's utility function is the discovery of an online shopper's preference function (modeled as an ID) via estimation of the ID's decision node coefficients using observations of the shopper's purchasing behavior (see Chajewska *et al.* 2001).

4.3.1   Agreement functions

4.3.1.1   Hellinger distance

Several of the agreement measures used in consistency analysis are functions of a probability-based measure of the distance between two probability distributions called the *Hellinger distance*. This distance in turn is defined in terms of a hybrid probability function, called the *probability density probability function* (PDPF) defined next.

Let **U** contain an ID's nodes (random variables) and let β parameterize this ID. Partition **U** into $\mathbf{U}^{(d)}$ and $\mathbf{U}^{(ac)}$ – the vectors of discrete and absolutely continuous chance nodes, respectively. The joint cumulative probability distribution function of $\mathbf{U}$ can be decomposed as $F_{\mathbf{U}}(\mathbf{u}) = F_{\mathbf{U}}^{(d)}(\mathbf{u}) + F_{\mathbf{U}}^{(ac)}(\mathbf{u})$ where $F_{\mathbf{U}}^{(d)}(\mathbf{u})$ is the purely discrete component – completely determined by the multivariate probability mass function, $P(\mathbf{U} = \mathbf{u})$, and $F_{\mathbf{U}}^{(ac)}(\mathbf{u})$, the purely absolutely continuous component – completely determined by the multivariate probability density function, $f_{\mathbf{U}}(\mathbf{u}) = \partial F_{\mathbf{U}}^{(ac)}(\mathbf{u}) / \partial \mathbf{U}$ (Koopmans 1969). Koopmans gives a hybrid of the probability mass function and probability density function called the PDPF that allows the computation of joint probabilities of $\mathbf{U}$ when that vector contains a mixture of discrete and continuous random variables. The PDPF is defined as

$$pf_{\mathbf{U}}(\mathbf{u}) \equiv \frac{\partial}{\partial \mathbf{U}^{(ac)}} P(\mathbf{U}^{(d)} = \mathbf{u}^{(d)}, \; \mathbf{U}^{(ac)} \leq \mathbf{u}^{(ac)}). \tag{4.2}$$

For a joint event, $\mathbf{u}^{(i)}$ described by an ID, this function is written as $pf_{\mathbf{U}^{(i)}|\boldsymbol{\beta}^{(i)}}(\mathbf{u}^{(i)})$ (see Haas 2001). This function is a generalization of a probability mass function and a probability density function and is necessary because IDs that make up the political–ecological system simulator may contain a mixture of continuously valued and discretely valued chance nodes.

In terms of the PDPF, the Hellinger distance between two probability distributions is

$$\Delta(\boldsymbol{\beta}_1, \boldsymbol{\beta}_2) \equiv \left[\int \left(\sqrt{pf_{\mathbf{U}|\boldsymbol{\beta}_1}(\mathbf{u})} - \sqrt{pf_{\mathbf{U}|\boldsymbol{\beta}_2}(\mathbf{u})} \right)^2 d\mathbf{u} \right]^{1/2} \tag{4.3}$$

(see Tamura and Boos 1986).

Minimum Hellinger distance estimation offers a very practical trade-off between robustness and efficiency (Patra *et al.* 2008, Pak 1996). Maximum likelihood estimation, on the other hand, has poor robustness properties (Patra *et al.* 2008). But the building of stochastic political–ecological models is in its infancy. Therefore, use of a robust statistical estimator to fit such models is seen to be crucial so that there is some defense against the consequences of model misspecification.

4.3.1.2   Data agreement functions

Call a time series of action–actor–target observations an *actions history* data set. Let $g_S^{(Grp)}(\boldsymbol{\beta})$ be the agreement between the sequence of group actions produced by the IntIDs model and the actions history data set. Let $g_S^{(Eco)}(\boldsymbol{\beta})$ be the agreement between the time series of mean values on ecosystem output variables computed by the ecosystem ID and the time series of observations on these ecosystem variables. For the entire IntIDs model, $g_S(\boldsymbol{\beta}) = g_S^{(Grp)}(\boldsymbol{\beta}) + g_S^{(Eco)}(\boldsymbol{\beta})$.

Agreement with actions history data The `Overall Goal Attainment` (`OGA`) node in a group ID represents the group's perceived utility of a proposed decision option. The combination of an action and a target that maximizes the expected value of this node is the out-combination (decision option) that the group implements.

Let $\mathbf{out}_i^{(obs)}(t_j)$ be the observed **output action–target** combination of group i at time t_j; $\mathbf{out}_i^{(opt)}(t_j)$ be the **output action–target** combination computed by the group i's ID at that time; and M_{ij} be unity if $\mathbf{out}_i^{(opt)}(t_j) = \mathbf{out}_i^{(obs)}(t_j)$ and zero otherwise. Then $g_S^{(Grp)}(\boldsymbol{\beta}^{(Grp)}) = \sum_{i=1}^m g_S^{(i)}(\boldsymbol{\beta}^{(i)})$ where

$$g_S^{(i)}(\boldsymbol{\beta}) \equiv \sum_{j=1}^T M_{i,j}. \qquad (4.4)$$

When such actions history data has been observed without error, c_H should be small, say $< .01$ because this data has no sampling variability and is generated by the ID's true parameter values. If c_H is set to exactly zero, consistency analysis becomes a minimum distance parameter estimation method (see Equation 4.1).

Agreement with ecosystem state data Say that a multivariate time series of ecosystem node values has been observed. For example, in the East African cheetah EMT, cheetah and herbivore abundances are observed over time. Let $\mathbf{u}_{obs}(t)$ be the vector of these values at time t. This vector constitutes a size 1 sample on the observable ecosystem ID nodes at t. For such a sample, the negative Hellinger distance is

$$-\left|1 - \sqrt{pf_{\mathbf{U}^{(Eco)}|\boldsymbol{\beta}^{(Eco)}}(\mathbf{u}_{obs}(t))}\right|$$

(see Lindsay 1994). Note that an implication of this definition is that no ergodicity assumption is made about this multivariate time series.

Let $g_S^{(Eco)}(\boldsymbol{\beta})$ be the sum of these negative Hellinger distances across all combinations of R regions and T time points.

If all q variables in the ecosystem model are discrete and each takes on N values,

$$g_S^{Eco}(uniform) = RT\left[\sqrt{1/N^q} - 1\right]$$

and hence can be used to identify a lower bound. For example, when $R = 5$, $T = 100$, $q = 10$, and $N = 100$, a lower bound for g_S^{Eco} is -500. When all variables in the ID are discrete, the upper bound is 0.

4.3.1.3 Hypothesis agreement function

Values of the PDPF under an ID's hypothesis distribution, $\mathbf{U}^{(i)}|\boldsymbol{\beta}_H^{(i)}$, and its $\mathbf{U}^{(i)}|\boldsymbol{\beta}^{(i)}$ distribution are approximated by first drawing a size n sample of design points from a multivariate uniform distribution on the ID's chance nodes, $\mathbf{u}_1, \ldots, \mathbf{u}_n$, followed by an estimate of $pf^{(i)}_{\mathbf{U}^{(i)}|\boldsymbol{\beta}}(\mathbf{u}_i)$ with an l nearest-neighbor, nonparametric density estimator due to Thompson and Tapia (1990, p. 179) at each of these design points. Using these estimates, the Hellinger distance between $\mathbf{U}^{(i)}|\boldsymbol{\beta}^{(i)}_H$ and $\mathbf{U}^{(i)}|\boldsymbol{\beta}^{(i)}$ is

$$\hat{\Delta}^{(i)}\left(\boldsymbol{\beta}^{(i)}, \boldsymbol{\beta}^{(i)}_H\right) \equiv \left\{ \sum_{j=1}^{n} \left[\sqrt{\hat{pf}_{\mathbf{U}^{(i)}|\boldsymbol{\beta}^{(i)}_H}(\mathbf{u}_j)} - \sqrt{\hat{pf}_{\mathbf{U}^{(i)}|\boldsymbol{\beta}^{(i)}}(\mathbf{u}_j)} \right]^2 \right\}^{1/2}. \qquad (4.5)$$

This use of density estimation to form an approximation to a likelihood function that has an intractable analytical form is called simulated maximum likelihood (SML) (see Mariano *et al.* 2000, Brandt and Santa-Clara 2002, Jank and Booth 2003, Jank 2006, and Park and Gupta 2009).

Here, the square root of the simulated likelihood function is used to approximate the Hellinger distance. Hence, when $c_H = 0$, consistency analysis is a simulated minimum Hellinger distance estimator. This estimator is shown by Takada (2009) to have favorable robustness and efficiency properties.

For the ith group ID,

$$g^{(i)}_H(\boldsymbol{\beta}^{(i)}) \equiv \sum \left\{ -\hat{\Delta}^{(i)}(\boldsymbol{\beta}^{(i)}, \boldsymbol{\beta}^{(i)}_H) \right\} \qquad (4.6)$$

where the sum is taken over all belief networks formed when the ID is conditioned on unique in-combination–out-combination pairs that arise as the ID encounters actions directed against itself (an in-combination) and considers the utility of a possible reaction (an out-combination) over the IntID's time interval.

These per-group measures of agreement are summed across all groups in the IntIDs model to yield

$$g^{(Grp)}_H(\boldsymbol{\beta}^{(Grp)}) \equiv \sum_{i=1}^{m} \left\{ g^{(i)}_H \right\}. \qquad (4.7)$$

And for the ecosystem ID,

$$g^{(Eco)}_H(\boldsymbol{\beta}) \equiv \sum \left\{ -\hat{\Delta}^{(Eco)}(\boldsymbol{\beta}^{(Eco)}, \boldsymbol{\beta}^{(Eco)}_H) \right\} \qquad (4.8)$$

where the sum is taken over all regions at the last time point only.

Then, for the entire IntIDs model the measure of agreement with the hypothesis parameter values is $g_H(\boldsymbol{\beta}) = g^{(Grp)}_H(\boldsymbol{\beta}^{(Grp)}) + g^{(Eco)}_H(\boldsymbol{\beta}^{(Eco)})$.

The ecosystem ID contains a system of stochastic differential equations (see Chapter 8). Because this system is a Markov process, the use of simulation to approximate the likelihood function en route to computing the Hellinger distance results in a consistent parameter estimator (Hurn *et al.* 2003).

4.3.1.4   Summary measures to assess agreement

After the maximization procedure is executed, summary measures are needed to aid the analysis of hypothesis and consistent parameter values. One such summary measure can be based on a distribution's mode as follows. First, let n_{dists} be the number of conditional distributions that define the ID. After the **maximize** consistency analysis step executes, there are n_{agree} of these distributions for which the distribution's mode is the same under both hypothesis parameter values and consistent parameter values. The ratio n_{agree}/n_{dists} is an informal measure of how much of the theory that drove the specification of $\boldsymbol{\beta}_H$ is in agreement with observations.

4.4   The MPEMP: definition and construction

4.4.1   Definition

Let $\mathbf{Q}(\boldsymbol{\beta})$ be a vector of ecosystem state quantities that are modeled by the simulator's ecosystem ID. For example, $\mathbf{Q}(.)$ could be cheetah abundance and herbivore abundance in the year 2060. Assume that a consistency analysis has produced a set of consistent parameter values contained in $\boldsymbol{\beta}_C$ and the hypothesis parameter vector has been updated to these values, that is, $\boldsymbol{\beta}_H = \boldsymbol{\beta}_C$. Using this updated $\boldsymbol{\beta}_H^{(Grp)}$, one way to quantify the concept of a practical ecosystem management plan described in the Introduction is to associate political feasibility with the value of $g_H^{(Grp)}(\boldsymbol{\beta}_{\text{MPEMP}})$ where $\boldsymbol{\beta}_{\text{MPEMP}}$ is a set of group ID parameter values that have been modified away from this updated $\boldsymbol{\beta}_H^{(Grp)}$ so that a desired ecosystem state is achieved by a sequence of group ID actions over a future time period. Here, a desired ecosystem state is defined to be a set of desired expected values for the components of $\mathbf{Q}(.)$. This set will be denoted by $\mathbf{q}_d$. If a solution is found, $E[\mathbf{Q}(\boldsymbol{\beta}_{\text{MPEMP}})] = \mathbf{q}_d$.

The idea is to find a set of minimal changes in group beliefs from those represented by $\boldsymbol{\beta}_H^{(Grp)}$ so that these groups change their behaviors enough to allow the ecosystem to respond in a desired manner. In other words, the MPEMP is the ecosystem management plan that emerges by finding group ID parameter values that result in a desired ecosystem state but that deviate minimally from $\boldsymbol{\beta}_H^{(Grp)}$. Formally, $\boldsymbol{\beta}_{\text{MPEMP}} = \arg\max_{\boldsymbol{\beta}^{(Grp)}} \left\{ g_H^{(Grp)}(\boldsymbol{\beta}^{(Grp)}) \right\}$ under the constraint that the ecosystem ID produces expected output values that are close to those in $\mathbf{q}_d$.

4.4.2   MPEMP construction procedure

First, define an *ecosystem damage utility function*, $f_{\text{ecodam}}(\boldsymbol{\beta}^{(Grp)})$, to be $\sum_{i,j,t} E[OGA_t^{(i)} | ecodam\text{-}action_{i,j}]$ where $OGA_t^{(i)}$ is group i's OGA node at time t, and *ecodam-action*$_{i,j}$ is group i's jth action that causes damage to the ecosystem.

The sum is over all groups that directly affect the ecosystem, all ecosystem-damaging actions by these groups, and all time points at which such actions are executed during a run of the simulator. One way to construct the MPEMP is with the following three-step procedure:

Step 1: Perform a consistency analysis with the current β_H and a political–ecological data set to find β_C. Update β_H to this β_C and then compute $\mathbf{q}_H = E[\mathbf{Q}(\beta_H)]$.

Step 2: Specify $\mathbf{q}_d$; for example, within the East African cheetah EMT, $\mathbf{q}_d$ might be 2000 cheetahs and 10 000 herbivores in 2060.

Step 3: Set $k = 1$ and execute the following three-step iterative optimization algorithm:

Step 3.1: Compute

$$\beta_k^{(Grp)} = \arg\min_{\beta^{(Grp)}} \left\{ \left[\frac{f_{\text{ecodam}}(\beta^{(Grp)})}{f_{\text{ecodam}}(\beta_H^{(Grp)})} \right] + d(\beta) \right\} \tag{4.9}$$

subject to

$$\frac{|g_H^{(Grp)}(\beta_H^{(Grp)}) - g_H^{(Grp)}(\beta^{(Grp)})|}{g_H^{(Grp)}(\beta_H^{(Grp)})} < .1k \tag{4.10}$$

where $d(\beta) = ||E[\mathbf{Q}(\beta)] - \mathbf{q}_d||/||\mathbf{q}_H - \mathbf{q}_d||$ and $\beta = (\beta^{(Grp)'}, \beta_H^{(Eco)'})'$.

Step 3.2: If $d(\beta_k) = 0$ or $d(\beta_{k-2}) = d(\beta_{k-1}) = d(\beta_k)$, set β_{MPEMP} to $\beta_k^{(Grp)}$ and stop. The MPEMP has been constructed.

Step 3.3: Set $k = k + 1$.

During the execution of the optimization algorithm, as ecosystem-damaging actions become less attractive to a group, they will not be executed and hence will not contribute to the sum that forms $f_{\text{ecodam}}(\beta^{(Grp)})$. Note that different sequences of group actions can lead to different values in the vector $E[\mathbf{Q}(\beta)]$. Further, utilities of such actions need to be computed under relevant in-combinations, that is, within the context of a run over the time interval to be managed.

The above optimization algorithm constructs the MPEMP by sequentially reducing the utility of executing ecosystem-damaging actions under the (gradually weakened) constraint of staying close to the distribution defined by the updated $\beta_H^{(Grp)}$. Because the smallest changes away from values contained in $\beta_H^{(Grp)}$ have been found that achieve the desired ecosystem goals, there are no other group behavior changes that require smaller changes in group belief systems before such behaviors change enough to achieve the desired ecosystem goals. Hence, the ecosystem management plan that is based on β_{MPEMP} is the most politically feasible.

To implement the MPEMP in the real world, group beliefs that correspond to parameters that have large differences between the updated $\beta_H^{(Grp)}$ and β_{MPEMP} need to be changed in the direction of the β_{MPEMP} value. Methods currently used that attempt to change people's perceptions and values (belief systems) include educational programs, workshops, and advertising.

If the needed degree of belief system change appears to be beyond available resources, less practical ecosystem management plans can be constructed by minimizing $f_{ecodam}(\beta^{(Grp)})$ over parameters that define groups for which beliefs can be realistically changed and leaving the parameters of all other groups at their $\beta_H^{(Grp)}$ values.

It is possible that the desired ecosystem state values cannot be achieved under any pattern of output actions issued by group IDs. This situation is indicated by $d_i \gg 0$ at the last iteration of the optimization algorithm.

4.5   The MPEMP for East African cheetah

Refer to Chapter 1 for background on the problem of cheetah conservation in East Africa.

4.5.1   Setup and computation

Say that a cheetah conservation goal is to have an expected cheetah abundance of 200 individuals in the Kenyan district of Tsavo 50 years hence, that is, in the year 2060. Say that only rural resident and pastoralist groups are to have their belief systems modified, with all other groups having their parameters held at their updated $\beta_H^{(Grp)}$ values.

Figure 4.1 shows the political–ecological data set used to find the consistent parameter values of the political–ecological system simulator. Chapter 2 contains a detailed explanation on how to interpret the actions history plot in this figure.

In the ecosystem variable plot of this figure, the ecosystem ID's variables Cheetah Fraction Detected, Herbivore Numbers, and Cheetah Non-Survey Sightings are displayed with lower-case alphabetic, upper-case alphabetic, and nonalphabetic characters, respectively. Each symbol in this plot is a region. Each ecosystem variable observation has been divided by its maximum value. These maximum values are 1.0, 993 732.0, and 45.0 for the above three variables, respectively. A line connects two observations on an ecosystem variable that were taken in the same region.

In order to show what is predicted to happen if group belief systems remain at their consistent values, Figure 4.2 shows this consistent model run over the years 2010 through 2060. For purposes of illustration, parameters governing the strength of belief held by Kenya rural residents in being arrested for poaching are set low.

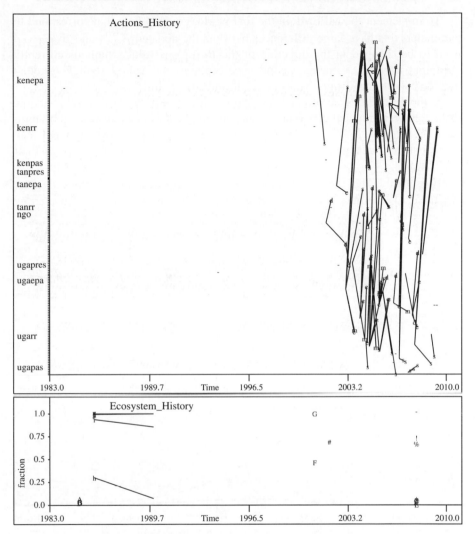

Figure 4.1   A political–ecological data set. The top plot contains the observed actions history used to fit the East African cheetah EMT simulator. The bottom plot contains the ecosystem observations used to fit this simulator's ecosystem ID. See the text for further explanation of these plots.

The MPEMP solution sets the values of these parameters so that the belief in arrest is high, with all other parameters remaining at their consistent values. Figure 4.3 shows this MPEMP solution over this same time interval.

Under the values in the updated $\beta_H^{(Grp)}$, the expected cheetah abundance in the year 2060 for this district is zero, but under β_{MPEMP} the cheetah population is viable. A comparison of Figures 4.2 and 4.3 reveals that the action *poach for cash*

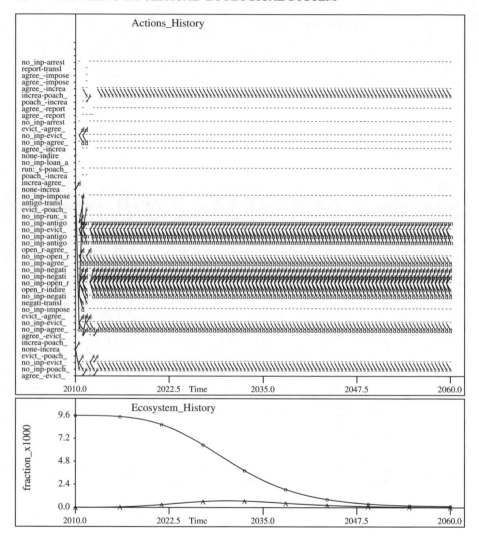

Figure 4.2   Simulator output over the time period of 2010 through 2060 using consistent parameter values. A pentad label consists of the string formed from the first six characters of the input action, a hyphen, and then the first six characters of the output action. In the ecosystem history plot, the 'o' curve is the mean of the variable Cheetah Fraction Detected *in the Tsavo region of Kenya, and the 'A' curve is the standard deviation of this variable.*

is not being executed in the MPEMP scenario. This is because the values contained in the updated $\beta_H^{(Grp)}$ that cause this action to be perceived as economically attractive with low risk of arrest are changed to values in β_{MPEMP} such that it is perceived by these groups to have a large risk of arrest.

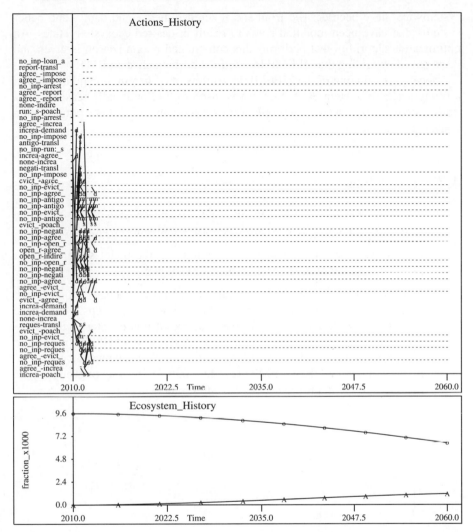

Figure 4.3   Simulator predictions for 2010–2060 using MPEMP parameter values. Compare to Figure 4.2.

4.6   Conclusions

A new definition of an optimal ecosystem management plan has been given in this chapter. This definition makes reaching ecosystem state goals the primary objective and is implemented as a constraint in the optimization algorithm. Changing group belief systems as little as possible, implemented as the objective function in the optimization algorithm, is treated as a secondary objective. This secondary objective is operationalized by maximizing the agreement between belief systems that have been derived from social science theory and data – and belief systems that have been modified so as to result in desired ecosystem states. An optimization algorithm that performs this constrained maximization is given and is exercised by finding the MPEMP for the East African cheetah EMT.

Because the finding of the MPEMP is demonstrated herein, it can be said that a freely available system exists that can be used to find politically feasible solutions to ecosystem management problems in the sense that predefined conservation goals are reached.

4.7   Exercises

1. Using **id**, find the MPEMP in the East African cheetah EMT under the restriction that the belief systems of the rural residents cannot be changed.

2. Repeat Exercise 1 but under the restriction that the belief systems of the president and the rural residents cannot be changed.

3. Find the best set of ecosystem state values that can be reached by any change in group beliefs. Do this by iterating the following two steps. First, make a small change to q_d as given in the example of this chapter in the direction of increased ecosystem health. Second, rerun **id**'s MPEMP procedure using this slightly modified set of desired ecosystem state values. Continue in this manner until the MPEMP procedure fails to find a set of group beliefs that leads to the improved set of ecosystem state values.

5

An open, web-based ecosystem management tool

5.1 Introduction

As outlined in Chapter 1, this author believes that there is a need for a publicly accessible EMT whose software and data are in the public domain. This chapter contains a description of such a tool and how it would be used by both ecosystem managers and observers.

The chapter's layout is as follows. First, an overview is given of the EMT's main components: the user interface, those political–ecological data sets relevant to the ecosystem being managed, and the analysis software. The software component is a web application dubbed **id** (for 'Influence Diagram'). **id** supports data acquisition, modeling, model fitting, model reliability assessment, and construction of model-based management plans. The language used to communicate with **id** is motivated and then completely specified. Lastly, a recommended sequence of transactions with the EMT is given that would support the analysis needs of the two main EMT user types: a manager with authority to implement a suite of actions that can affect the ecosystem, and a passive observer of an ecosystem that is being managed by others.

5.2 Components of a politically realistic EMT

5.2.1 User interface

To ensure ease of use, the user interface is a critical aspect of the EMT. A user possessing only modest statistical training should be able to ask the EMT for a

Improving Natural Resource Management: Ecological and Political Models   Timothy C. Haas
© 2011 John Wiley & Sons, Ltd

particular ecosystem analysis. In addition, the EMT should be able to generate at least part of the publication quality report that the user is almost always ultimately interested in as a final product of the analysis.

Many ecosystem data analyses are incremental and sequential – a preliminary analysis is performed and then a sequence of analyses are executed, each only minimally different from the original analysis. Because of this sequential nature of performing ecosystem analyses, it is more convenient to have a natural language interface to the EMT rather than a so-called graphical user interface (GUI). Therefore, users interact with the EMT through a language that has been developed to describe, estimate, and evaluate an ID and/or an IntIDs model.

Because the computing system component of the EMT is called **id**, this language is referred to as the '**id** language.' Once learned, the **id** language allows the user to quickly formulate the necessary instructions to the EMT for both performing the desired analysis and generating all necessary figures and tables for the report on that analysis.

5.2.2   Data collection

Political and ecological data would need to be collected at regular intervals on a permanent basis.

5.2.3   Actions history plots

Reports generated by this computing system would typically include plots of actions histories. An ecosystem manager would want to see the entire actions history plot and would view it on a screen using a horizontal scrollbar to move across time so that, in practice, the plot would not be compressed in the time dimension. Because this manager would be concerned about inter-country action–reaction pairs, it would be important to not excessively compress the time dimension so that such action–reaction pairs (and more lengthy interactions) could be easily discerned.

5.2.4   EMT website architecture

The architecture of this website, displayed in Figure 5.1, is as follows. There are three main directories (folders): `data_sets`, `input_files`, and `report`. The `data_sets` directory has `political` and `ecosystem` sub-directories; the `input_files` directory has no sub-directories; and the **report** directory has sub-directories for model reliability, model output displays (`out-files`), and the MPEMP. The model reliability sub-directory has its own sub-directories: model-fit-to-data statistics (`ca_statistics`), one-step-ahead prediction error

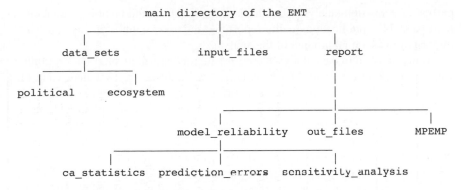

Figure 5.1   Directory tree of a typical EMT website.

rates (prediction_errors), and a sensitivity analysis of the model's parameters (sensitivity_analysis).

5.3   id language and software system

All hardware and operating system mechanics for running **id** are described at www4.uwm.edu/people/haas/idusers. The language presented below is a fully functional language for performing statistical analyses needed to support ecosystem management decision making.

5.3.1   Language overview

Following Johnson (1994), the **id** language is defined by a hierarchical vocabulary of main words, qualifiers of those words, and a set of *n*-ary assembly relations.

There are only four main words: influence_diagram, node, context, and report. The latter three words have several qualifiers and all have several *n*-ary assembly relations (see Tables 5.1–5.4). Qualifiers specify particular types of a main word's entity, and relations create a mapping between user-created inputs and outputs that are associated with the qualified main word.

The user prepares an input file consisting of statements in this language. Each group of records starts with one of the main words followed by some number of qualifiers and/or relations. This language, similar to the one proposed by Lubinsky (1990), is intended to be a high-level, nonprogramming language for expressing statistical models and analyses that persons with only modest technical training can use to express political–ecological systems and find plans to manage the associated ecosystem.

The entire vocabulary of this language appears in Tables 5.1–5.4. In these tables and the following descriptions of the constituent relations, a Document Archive

Table 5.1 Qualifiers and relations for the **id** language words influence_diagram and node. Arguments to relations are italicized.

Word-level relations	Qualifiers	Qualifier relations
	influence_diagram	
idfile(*id_file_name, plotting_string, country_name*)		
	node	
names(*node_name short_name*)	Determ_Contin	parameters(*n parm-1 ... parm-n*)
values(*n value-1 ... value-n*)	Determ_Decision	parameters(*n parm-1 ... parm-n*)
parents(*n short_name-1 ... short_name-n*)	Determ_Discrete	
memory(*short_name*)	Determ_Frctn	parameters(*n parm-1 ... parm-n*)
	Determ_Labeled	parameters(*n parm-1 ... parm-n*)
	Determ_Linear	parameters(*n parm-1 ... parm-n*)
	Determ_Loss	parameters(*n parm-1 ... parm-n*)
	Determ_Root	
	Determ_Thrshld	parameters(*n parm-1 ... parm-n*)
	Discrete	
	Gamma	parameters(*n parm-1 ... parm-n*)
	GLOMAP	parameters(*n parm-1 ... parm-n*)
	Logit	parameters(*n parm-1 ... parm-n*)
	Lognormal	parameters(*n parm-1 ... parm-n*)
	LOMAP	parameters(*n parm-1 ... parm-n*)
	Normal	parameters(*n parm-1 ... parm-n*)
	SDE	parameters(*n parm-1 ... parm-n*)

Table 5.2 Qualifiers and relations for the **id** language word `context`.

Qualifier	Qualifier relations
`files`	`parameter_files`(*old/new hyp_file_name est_file_name init_file_name MPEMP_file_name*)
	`boundaries_file`(*file_name x-is-longitude (true, false)*)
	`data_file`(*file_name*)
	`sites_file`(*file_name*)
	`actions_file`(*file_name*)
	`output_files`(*graph_file_name.ps report_file_name.html parameter_estimates_file_name*)
`settings`	`javaspaces_identification`(*jini_URL space_name*)
	`real-world_phenomenon_type`(*type*)
	`geographic_window`(*min_long max_long min_lat max_lat*)
	`ecosystem_nodes`(*n label-1 ... label-n*)
	`inputs`(*n (node-1 value) ... (node-n value) (Time sdebegintime tmin tmax) (node-i _all_))*
	`initial_action`(*actor action n target-1 ... target-n*)

Number (DAN) is a unique identification number assigned by the database builders to a news article for the purposes of building a group actions database. For details of how such a database is built, see Chapter 9.

5.3.2   id language file example

Figure 5.2 contains an **id** language file for modeling the deposition of NO_3 through precipitation with `LOMAP` spatial stochastic random variables.

5.3.3   Descriptions of `influence_diagram` and `node`

5.3.3.1   `influence_diagram`

This main word specifies an ID language file that, in turn, describes one of the component IDs of an IntIDs model. Its sole relation includes strings for plot labels and the country that the ID is associated with.

5.3.3.2   `node`

A node represents either a random (stochastic) variable or a deterministic variable. These variables may be observable or latent. Random variables are called *chance nodes*. The ID structure contains LISREL (Koster 1996) models which, in turn, contain multivariate linear models. Such models can be represented in **id** with `Determ_Linear` nodes.

Table 5.3 Qualifiers and relations for the **id** language word `report`.

Qualifier	Qualifier relations
`prepare_data`	`convert_data`(*lng-lat_file_name x-y_file_name*)
	`convert_eos_data`(*eos_lng-lat_file_name x-y_file_name file_min_longitude file_max_longitude file_min_latitude file_max_latitude time #rows #cols min_val max_val background_value variable_name*)
	`convert_boundary`(*n lng-lat_file_name-1 ... lng-lat_file_name-n x-y_file_name #rows #cols*)
	`convert_label-grid`(*label_file_name surface_file_name*)
	`grid`(*#rows #cols PostScript_file_name #rows #cols*)
	`prepare_raw_group_data`(*raw_data_file_name actions_history_data_file_name check_level*)
	`gis_tools`(*latitude-longitude_file*)
	`parse_stories`(*story_file_prefix story_file_postfix #story_files starting_DAN group_acronyms_file*)
`describe_data`	`compare_samples`(*n (node-a-1 node-b-1) ... (node-a-n node-b-n)*)
`estimate`	`estimate_nodes`(*begin_time end_time m (idname-1 n_1 node-1_1 ... node-n_1) ... (idname-m n_m node-1_m ... node-n_m) c_H #MC_realizations*)
	`fit_logits`(*n node-1 ... node-n*)
	`prediction_skill`(*begin_time end_time m (idname-1 n_1 node-1_1 ... node-n_1) ... (idname-m n_m node-1_m ... node-n_m) c_H #MC_realizations*)
	`cross-validation`(*min_time max_time pred_option pred_time*)
	(If there is no 'Time' node, the time values to the cross-validation relation are ignored.)
	`find_mpemp`(*begin_time end_time m (idname-1 n_1 node-1_1 ... node-n_1) ... (idname-m n_m node-1_m ... node-n_m) c_H #MC_realizations*)

Semiparametric and parametric spatio-temporal models are represented with LOMAP or GLOMAP nodes (see Haas 2002). Hence, the collection of node statements in an ID input file defines the stochastic model of the relationship between the dependent nodes and independent nodes.

It is planned to include nonparametric models such as neural networks, classification and regression trees and k-nearest-neighbor classifiers by creating the qualifiers NN, Tree, and KNN, respectively.

```
Nitrate, Precipitation example
of bivariate spatial prediction.
node Xcoord                              x          Determ_Decision
   values(1 xcoord)


node Ycoord                              y          Determ_Decision
   values(1 ycoord)


node no3                                no3         LOMAP
   parents(2 x y)
   values(1 no3)
   parameters(5 trnsfrm kmax vmodel itrend f_c)


node ppt                                ppt         LOMAP
   parents(2 x y)
   values(1 ppt)
   parameters(5 trnsfrm kmax vmodel itrend f_c)


context
   real-world_phenomenon_type(ecosystem)
   data_file(spatialpred.dat)
   parameter_files(old spatialpred-hyp.par spatialpred-est.par)
   boundaries_file(spatialpred.bln false)
   sites_file(sites.dat)
   output_files(spatialpred.ps spatialpred.html spatialpred.est)


report estimate
   cross-validation


report evaluate
   surface(no3 0. 0. 0. 0. 2 9 1)


report display
   plot_surface(sill.sgd fig2.ps)
   display_intervals(5 0. .15 .15 .3 .3 .45 .45 .6 .6 .75)
   interval_labels(5 '.00-.15' '.15-.30' '.30-.45' '.45-.60' '.60-.75')


report prepare_data
   convert_boundary(1 us.lgt us.bln 60. 130. 25. 52. 5 5)
```

*Figure 5.2   **id** language file for modeling NO₃ deposition through precipitation.*

5.3.3.3   Word-level relations

These relations are not associated with any qualifier.

names : gives the node's full name and a short abbreviation. Express node names
   with short phrases and use "_" to connect words. For the short name, use either
   an acronym or other abbreviation of the node name that is suitably short for
   graphical displays, say 1–5 characters.
      Arguments: *node_name, short_name.*

values : gives the values that the node can take on.
      Arguments: *number of values (n), value-1, ... value-n.*

Table 5.4   Qualifiers and relations for the **id** language word: `report`, continued.

Qualifier	Qualifier Relations
`evaluate`	`id_interactions`(*begin_time end_time #MC_realizations per ID*) `optimal_decision`(*#MC_realizations*) `evaluate_nodes`(*n node-1 ... node-n #MC_realizations*) `surface`(*short_name min_data-read_time max_data-read_time begin_pred_window_time end_pred_window_time pred_option #rows #cols grid_type*) `volume`(*short_name min_time max_time pred_time pred_option #rows #MC_realizations*)
`display`	`plot_graph`() `label_regions`(*PostScript_file*) `overlay`(*surface_file PostScript_file*) `display_intervals`(*n (low-1 high-1) ... (low-n high-n)*) `interval_labels`(*n label-1 ... label-n*) `map_levels`(*n original_level_name-1 new-level-1 ... original_level_name-n new-level-n*) `plot_surface`(*surface_file PostScript_file*) `plot_actions_history`(*plot_type actions_history_file*)
`sensitivity_ analysis`	`assess_nodes`(*begin_time end_time m (idname-1 n_1 node-1_1 ... node-n_1) ... (idname-m n_m node-1_m ... node-n_m) #MC_realizations*)
`mc_hypothesis_ test`	`test_nodes`(*begin_time end_time m (idname n node-1 ... node-n) #MC_realizations*)

`parents`:   gives the parents of the node.
   Arguments: *number of parents (n), parent-1, ..., parent-n.*

`memory`:   gives the node whose distribution at the previous time step is to be used by this node.
   Argument: *short_name.*

5.3.3.4   Qualifiers

`Determ_Contin`: a node that takes on a single continuous value for each combination of the values of its parents.
   There is no `parameter` relation for this qualifier.

Determ_Decision: a node with a user-specified list of values that are the different decision options being considered.
There is no parameter relation for this qualifier.

Determ_Discrete: a node that takes on a single discrete value for each combination of the values of its parents.
There is no parameter relation for this qualifier.

Determ_Frctn: a node whose value is 0 if its sole parent is below a lower threshold, between 0 and 1 if its parent is between the lower and upper thresholds, and 1 if the parent is above the upper threshold.
Arguments to parameter relation: *lower_threshold* and *upper_threshold*.

Determ_Labeled: deterministic node whose values are labels indexed by the numerical value of its parent.
Arguments to parameter relation: *label-1, ..., label-m* where the parent has *m* values.

Determ_Linear: a node that is a deterministic, linear function of its parent: $\beta_1 + \beta_2 \times$ parent value. Note that if a spatio-temporal covariance structure is desired, do not use this node type instead, use LOMAP/GLOMAP node type having the desired covariates and/or qualitative nodes as parents.
Arguments to parameter relation: β_1 and β_2.

Determ_Loss: a deterministic quadratic loss function centered at β_0 and scaled by β_1.
Arguments to parameter relation: β_0 and β_1.

Determ_Root: a deterministic root node (no parents).
There is no parameter relation for this qualifier.

Determ_Thrshld: a node whose value is 0 until the value of a parent node rises above a parameter threshold value, that is, 0 if parent value is less than *threshold*, 1 otherwise.
Arguments to parameter relation: *threshold*.

Discrete: the simple discrete chance node with a small (< 5) number of values.
There is no parameter relation for this qualifier.

Gamma: the gamma chance node.
Arguments to parameter relation: *shape* and *scale*.

Logit: The cumulative logit model. A distribution is specified in the .par file by listing the desired conditional distributions. This is accomplished by specifying the node's distribution at three values of the parent node. **id** finds parameter values that cause the model to match these specified conditional distributions. This search is started at all coefficients set to zero. This mechanism allows the user to specify a logit model without having to express a desired set of logit probabilities in terms of corresponding coefficient values.
This model is: $logit j = \ln[P(Y \le j)/P(Y > j)] = \alpha_j + f(\boldsymbol{\beta}, \mathbf{X}), j = 1, \ldots, J - 1$, where $\mathbf{X}$ is a vector of m parent values, and $\boldsymbol{\beta}$ is the vector of corresponding coefficients. For $J = 3$, the probabilities for each level

of Y are: $p_1 = \exp(logit1)/(1 + \exp(logit1))$ and $p_2 = (\exp(logit2) - p_1(1 + \exp(logit2)))/(1 + \exp(logit2))$. Functions $f()$ of the parent node are coded into the method `Beliefs.complogits_()`.

Arguments to `parameter` relation: $\alpha_1, \ldots, \alpha_J, \beta_1, \ldots \beta_m$.

`LOMAP`, `GLOMAP`: multivariate spatio-temporal stochastic processes. `LOMAP` uses a semiparametric moving-cylinder model for the trend along with local covariance structure models, while `GLOMAP` uses a kernel-weighted sum of `LOMAP` models to provide a global model of the spatio-temporal process's trend and covariance structure.

Arguments to `parameter` relation:

ttype: the value 1 indicates one-time-step-ahead forecasting; 2 indicates spatio-temporal interpolation.

kmax: number of spatial lags for the spatial covariogram.

kmaxt: number of temporal lags for the temporal covariogram.

vmodel: semivariogram model specification: 0 gives a nugget-only semivariogram; 1 a spherical, and 2 an exponential one.

itrend: order of the spatio-temporal trend polynomial: 4 for a zero-order polynomial; 5 for a first-order polynomial; and 6 for a second-order polynomial.

s_frac: fraction of monitoring sites (spatial locations) to use in a `LOMAP` prediction cylinder. Referred to as f_c in Haas (1995).

t_frac: temporal length of a `LOMAP` prediction cylinder.

trnsfrm: the value 0 produces no transformation to the residual process, and 1 produces a transformation (see Haas 2002).

nmstmixcntr: number of GLOMAP components (GLOMAP only).

Explanation of parameter terms: The following is based on Haas (1995). The term *prediction* will refer to inference on random quantities at any location and time, and the term *estimation* will refer to inference on fixed but unknown parameters.

Let the spatial coordinates of locations in the spatio-temporal space be given by (x, y) and the temporal coordinate by t. A spatio-temporal location is designated by $\mathbf{x} = (x, y, t)'$, and n is the total number of spatio-temporal observations. Let $f_c \in (0, 1)$ be the fraction of n that is used for a prediction. Define $n_c \equiv nf_c$ to be the number of observations used to calculate the prediction at $\mathbf{x}_0$. Call the spatio-temporal space that holds the n_c observations used to predict the process at $\mathbf{x}_0$ the *prediction cylinder*.

The cylinder's n_c observations are found as follows.

Step 1: Let t_{earliest} and t_{latest} be the time of the earliest and latest observation in the data set, respectively. The temporal range of the cylinder is fixed at a user-selected value, $m_T \leqslant t_{\text{latest}} - t_{\text{earliest}}$. The cylinder's temporal interval is $[t_L, t_U]$ where $m_T = t_U - t_L$. The upper limit, t_U, is defined to be $\min\{t_{\text{latest}}, t_0 + m_T/2\}$. The lower limit, t_L, equals $\max\{t_{\text{earliest}}, t_U - m_T\}$. It is assumed that m_T is large enough so that $n_c < n_I$.

Step 2: The n_I observations found in step 1 are sorted on the primary sort key of $\| (x_0, y_0)' - (x, y)' \|$ and on the secondary sort key of $|t_0 - t|$, that is, the sites are sorted according to their spatial distance from $(x_0, y_0)'$ and all observations taken at a particular site are sorted by their temporal distance from t_0. Let this list of sorted observations be numbered 1 through n_I.

Step 3: The cylinder's observation set is defined to be the first n_c of these sorted observations.

`Normal`, `Lognormal`: the normal and lognormal chance nodes, respectively. Arguments to `parameter` relation: μ and σ^2.

`SDE`: In this version of **id**, if a system of SDEs is part of the model, the mathematical forms of the trend and diffusion matrices need to be programmed in the `a_()` and `b_()` methods, respectively, in the `Java` source file, `Sdesol.java`. In the source distribution of **id**, the cheetah viability SDE system is coded in this file. Specifically, the four SDEs on pages 113 through 115 of Haas (2001) are coded within `Sdesol.java` as follows:

```
static double a_(double nodevals[][][], double alpha[], double beta[],
                int k, double tn, double yn[], int dst) {
// Evaluates the trend vector.
int i;
double val = 0., trend, y;
if (k == 1) {
   y = yn[k - 1];
   val = y * (alpha[0] - y);
} else if (k == 2 || k == 3) {
   y = 2. * yn[k - 1] - 1.;
   val = -.5 * (alpha[k - 1] + beta[k   1] * beta[k - 1] * y) *
      (1. - y * y);
} else if (k == 4) {
   trend = yn[1] * (1. - Math.pow(alpha[3], alpha[4] * yn[3])) * yn[3];
   trend -= yn[2] * yn[3];
   // Get the current Carrying Capacity node value.
   i = Getmodl.getndnm_("CarCap");
   alpha[5] = nodevals[Beliefs.locnm][1][i - 1];
   trend -= (yn[1] - yn[2]) * yn[3] * yn[3] / alpha[5];
   val = trend;
}
return val;
}


static double b_(double beta[], int i, int j, double tn, double yn[],
                int dst) {
// Evaluates the (i, j)^th component of the diffusion matrix.
double val = 0., y;
if (i != j) {
   val = 0.;
} else if (i == 1) {
   val = beta[0] * yn[0];
} else if (i == 2 || i == 3) {
   y = 2. * yn[i - 1] - 1.;
   val = beta[i - 1] * (1. - y * y);
```

```
} else if (i == 4) {
   val = beta[3];
}
return val;
}
```

To solve a different set of SDEs, modify `Sdesol.java` as necessary and then compile Sdesol.java to rebuild the JAVA class file, `Sdesol.class`.

Arguments to `parameter` relation: SDE dependent.

5.3.4   Description of `context`

This word sets the context for the report's analysis. Its qualifiers are as follows.

`files`: gives names of various input and output files needed for the analysis. Relations for this qualifier:

`parameter_files`: gives the file names of the hypothesis, initial values, and estimated (consistent) parameter files that define the ID's nodes. The parameter values associated with the model defined in a *name.id* file are listed in the parameters file which has the file extension `.par`. "name" is any alphanumeric string acceptable to the user's file system. The user is responsible for creating separate parameter files to hold hypothesis and estimated values, respectively. A suggested naming convention is *name-hyp.par*, *name-init.par*, and *name-est.par*. **id** writes analysis results to the report file.

The results of a run to find the Most Practical Ecosystem Management Plan (MPEMP) are written to *MPEMP_file_name*.

When `LOMAP/GLOMAP` node types are present, semivariogram parameter estimates are written to the file *name.est*.

`boundaries_file`: gives the file name containing spatial boundary files.

`sites_file`: gives the file name of the monitoring site locations.

`data_file`: gives the data file name. This file contains non actions history observations. The form of this file name is *name.dat*. It consists of observations on one or more of the ID's nodes and is organized as stacked sets of records, one stack for each observed chance node. Say there are n_i observations on the ith observed chance node. The records for this node are:

1. Record 1: A sequence of deterministic node short names plus the observed chance node's short name. These short names can be in any order but must be separated by at least one space.

2. Record 2: The word 'begin.'

3. Records 3 through $2 + n_i$: values on the nodes listed in record 1 and in the same order as the record 1 sequence.

4. Record $3 + n_i$: the word 'end.'

This data file form allows chance nodes to be irregularly and noncoincidentally observed in space and time.

`output_files`: gives file names for the report file and the file containing selected parameter estimates.

`settings`: provides information needed to narrow the analysis to a particular situation under study.

Relations for this qualifier:

`real-world_phenomenon_type`: currently, *president*, *EPA*, *rural_resident*, *pastoralist*, *ngo*, and *ecosystem* are recognized.

`geographic_window`: gives a rectangle in geographic (decimal) units for the analysis. Any points in a conversion activity that are outside this window are ignored. The bounding box defined by the `convert_boundary` relation over rides these values.

`initial_actions`: gives the actor and output action that initiates an IntID simulation.

`inputs`: specifies the values of the conditioning nodes. If the value of a discrete `inputs` node is the reserved keyword `_all_`, then each value of this node in turn is used as the conditioning value. Use this capability to compute model outputs over each and every region in a spatial data set. Such a node needs to be the first node to appear in this list. For the special case of the node 'Time,' specify the beginning time for an SDE solution, the minimum conditioning time, and the maximum conditioning time with:

'Time' *sdebegintime tmin tmax*.

If no SDEs are in the ID, *sdebegintime* is ignored. For a spatial-only LOMAP/GLOMAP node, computations will be performed for each unique time value in the sample between these minimum and maximum values. For a spatiotemporal LOMAP/GLOMAP node, a single computation using all observations between these two values will be performed.

`ecosystem_nodes`: specifies the node names that are the output nodes of the ecosystem ID.

5.3.5   Description of `report`

This word specifies the details of the report's analysis. Qualifiers are as follows.

`prepare_data`: prepares data files for subsequent analyses.

Relations for this qualifier:

`convert_data`: converts a longitude–latitude data file to an x, y data file.

`convert_eos_data`: converts an Earth Observation System (EOS) longitude–latitude data file to an x, y data file. The arguments *file_min_longitude*, *file_max_longitude*, *file_min_latitude*, *file_max_latitude*, *#rows*,

#cols, min_val, max_val, and *background_val* can be found in the header (.hdr) file that **ModisTool** creates when it reads the .hdf file obtained from the EOS data center. **ModisTool** can be downloaded for free from the EROS Data Center at http://lpdaac.usgs.gov/tools/modis/register.asp. In this header file, use NLINES for the *#rows* and NSAM-PLES for the *#cols.* If DATA_TYPE is not INT16, the Java source code file Eosdata.java will need to be modified accordingly. Running **ModisTool** is largely self-explanatory but be careful to select a geographic projection and a raw binary file for the output.

convert_boundary: converts a longitude–latitude boundary file to an *x, y* boundary file. This relation also writes a *#rows*-by-*#cols* longitude–latitude grid that overlays the bounded region. The boundaries and grid are plotted in the PostScript file 'lng-latgrd.eps.'

prepare_raw_group_data: reads a data file of group actions and creates an actions history file with actions translated into Ecosystem Management Actions Taxonomy (EMAT) categories (see Chapter 9). The relation's param-eter, *check_level*, takes on the values 'lowcheck' or 'highcheck.' The latter provides more complete printing while the file is being translated.

gis_tools: starts an interactive session to perform GIS operations. Below, these operations or *tools* will be referred to as **id**'s *GIS tools*. When **id** is run with this relation, menus allow the user to perform the following tasks:

1. Display geographic images, for example, maps.

2. Perform on-screen digitizing. Having this capability relieves the EMT-maintaining organization from having to purchase and maintain a digitizing tablet. This capability functions by having the user first enter the latitude–longitude coordinates of at least three points on the image. Then, **id** computes the minimum-error transformation between the image and the latitude–longitude coordinate systems. Then, the user is free to use the mouse to indicate points on a path that represents either a region of the image or a surface of constant (user-entered) value.

3. Convert a data set expressed in image coordinates to one that is expressed in latitude–longitude coordinates.

Although these GIS tools are available in other free or lease-only software packages, they have been incorporated into **id** so that a user needs to learn only one, free software system to perform ecosystem management analyses. See the conclusions of this chapter for further discussion of this development goal.

parse_stories: given a list of HTML-based story files, parses each story to create an EMAT actions entry. Execute the following steps to create EMAT entries:

1. Create a Google account and then create news alerts for a set of desired keyword phrases. Have these alerts sent to a mailbox.

2. Read each alert email. If a story seems relevant, open the story's link and write it to a file as an HTML-only file type ('webpage, HTML only' in Internet Explorer. Use a file name of s*n*.htm where *n* is a number, for example, s1.htm s2.htm etc.

3. Prepare the **id** input file by listing the files to be read by entering the file name prefix, postfix, the number of files to be read, the starting DAN, and the file containing group names in the vernacular, for example,

parse_stories(s htm 10 866 eastafgroups.dat)

to read 10 stories contained in the files s1.htm s2.htm ... s10.htm with actions being given DAN values starting with the number 866.

4. Add the action entries written to the output file (usually shell.out) to the actions data base, for example, eastafacts.dat.

5. The collection of news stories with all HTML tags removed will be contained in a file named dans*n*to*m*.dat where *n* is the starting DAN and *m* is the ending DAN in the file, respectively. In this file, each news story has a header line of the form

STORY: ———— start DAN = *n* end DAN = *m* ————-

Stories for which the relation's parsing algorithm failed are written to the file parsefailed.dat.

`describe_data`: requests descriptive statistics to be computed for each node in the data file. **id** output includes summary statistics on each such node and output files that support graphical displays of the data.
Relations for this qualifier:

`compare_samples`: compares *n* pairs of samples taken on the listed pairs of variables. Bootstrap tests for differences in median, interquartile range (IQR), skew, and kurtosis are reported.

`estimate`: requests parameter estimation via consistency analysis (see Chapters 4 and 11). Detailed statistical fit and assumption–satisfaction diagnostics are written to the report file.
Relations for this qualifier:

`fit_logits`: gives the logit model nodes for which logit parameters are to be found that result in logit probabilities matching as close as possible the conditional probabilities listed in the .par file.

`estimate_nodes`: gives the nodes whose parameters are to be estimated. All other parameters in the ID are held at their initial values (contained in *init-file_name*) during the estimation procedure. The *# of MC_realizations* parameter will need to be set high enough so that small changes to model parameters during the optimization algorithm cause detectable changes in the objective function.

Direct search is used to find parameter estimates and hence can be computationally expensive. See Appendix B for a description of a procedure used in **id** that speeds up this computation when a cluster of computers is available.

When an ID is to be evaluated in **id**, MC simulation is used in lieu of exact summation of conditional probability values since this latter method is known to be NP-hard (see Cooper 1987).

There are ways to verify that the *# of MC realizations* parameter has been set to a large enough value. The first method is to try progressively larger values until changes to the parameters of the node that is furthest from the node whose parameters are being fitted cause changes in the objective function. The second method is as follows. Let $m = $ *# of MC_realizations*. Every joint event in the ID can be represented via the *recursive factorization*. For example, in a three-node ID where node 3 is a root node, node 2 is influenced by node 3, and node 1 is influenced by nodes 3 and 2, the joint event $\{x_1, x_2, x_3\}$ can be written as

$$P(X_1 = x_1, X_2 = x_2, X_3 = x_3) = P(X_1 = x_1 | x_2, x_3) P(X_2 = x_2, X_3 = x_3)$$
$$= P(X_1 = x_1 | x_2, x_3) P(X_2 = x_2 | x_3) P(X_3 = x_3).$$

With some programming, the method `Jntprb.jntprb_()` contained in the source code file `Jntprb.java` can be used to find the joint event with the smallest probability. Let this smallest probability be p_{je}. Then, to see on average at least one realization of the joint event during the simulation, $m \geq 1/p_{je}$. For example, in the above example, if all conditional probabilities are .01, m needs to be greater than or equal to $1/.01^3$ or 10^6. If all of the probabilities are instead .1, m would need to be greater than or equal to 10^3.

`evaluate`: requests evaluation of an ID or an IntIDs model. If LOMAP/GLOMAP nodes are part of the ID, the model's parameters are estimated before the model is evaluated.

Relations for this qualifier:

`id_interactions`: **id** supports dynamic models of interacting IDs (an IntIDs model). Each ID is solved for its optimal decision from *begin_time* to *end_time*. Within this interval, time is incremented by an internal time step and then all IDs are solved again while taking into account the optimal decisions of other IDs that were computed one time step previously. **id** requires a master **id** language file that consists of `influence_diagram` words that specify the **id** language files of the constituent IDs.

`evaluate_nodes`: computes output on these nodes only.

`prediction_skill`: gives the nodes whose parameters are to be estimated – all other parameters in the ID are held at their hypothesis values during the estimation procedure. Then, one-time-step-ahead predictions are computed beginning at *firsttme*. The root mean squared prediction error (RMSPE) is computed for these true forecast errors (i.e., not cross-validation).

cross-validation: performs cross-validation on the observations in the data file from *min_time* to *max_time*.

surface: computes a surface of the dependent node via LOMAP predictions over the points of a grid enclosed by the boundary file. Setting *grid_type* to 1 produces a rectangular grid, and a setting of 2 produces a hexagonal grid.

If *pred_option* = 1, one-step-ahead prediction is performed, that is, model fitting and prediction are based on only the data that is one step earlier in time than the time specified by *pred_time*. If *pred_option* = 2, observations that are within the spatio-temporal cylinder and earlier, concurrent, or later than the prediction time are used (temporal interpolation).

volume: computes the total volume under a spatial surface. Currently, only a LOMAP-generated surface is available. This relation also computes an MC-based standard error of the volume estimate. The *pred_option* parameter is defined as above.

sensitivity_analysis: performs a sensitivity analysis.
Relations for this qualifier:

assess_nodes: gives within-ID lists of nodes whose parameters will be the subjects of the sensitivity analysis.

mc_hypothesis_test: performs an MC hypothesis test.
Relations for this Qualifier:

test_nodes: specifies nodes to be tested (*begin_time end_time m* (*idname n node*-1 ... *node-n*) *#MC_realizations*).

display: constructs graphical displays.
Relations for this qualifier:

plot_graph: creates a PostScript file of the ID's graph after computing node positions that minimize the number of link crossings.

label_regions: creates a PostScript file of all region boundaries and region labels.

overlay: computes and displays the fraction of each value of a discretely valued spatial node that is contained within each region.

display_intervals: defines a grayscale using the given list of intervals.

interval_labels: uses the given list of strings to label the grayscale legend.

plot_surface: creates a grayscale display of a surface.

plot_actions_history: creates a plot of an actions history. Arguments are *plot_type* and *actions_history_file*. An actions history file has two record types, the first being Record Type 1: Actions record. The format for this record is: 'action' *time, actor, action,* 'one,' 'two,' or 'three' (indicating ONE, TWO, or THREE targets), a comma, and then a comma-delimited list of targets.

Example:

```
2002.25 Kenya_rural_residents clear_new_land
one,ecosystem
```

This procedure is necessary because target node values can be of the variety

```
two,Kenya_rural_residents,Kenya_pastoralists
```

which is a single string but indicates two targets.

The second type is Record Type 2: Pentad record. The format of this record is: 'pentad' *output_action_identifier input_action_identifier output_action_time input_action input_actor DM-group output_action target.*

5.3.6 Structure of id language file

id uses one file to define an ID, called the **id** language file. This file contains all **id** language statements and is identified by its file extension `.id`. The structure of this file is as follows.

First section: all `node` statements to define the ID.

Second section: all `context` statements.

Third section: groups of `report` statements wherein each group begins with the main word, `report`.

id detects that a group of `report` qualifiers has ended when either the end of file is reached or another `report` statement is encountered. Only one `report` group is executed when **id** is run. Other analyses, described by other groups of `report` statements, can be stored below the group to be executed. Use of this feature allows all analyses to be maintained in one file.

In any of the relations of this language, if a character string is a number, enclose it in single right quotes, for example,

```
interval_labels(2 '1.5-2.5' '2.5-3.5').
```

5.3.7 Structure of a surface file

This file type is referred to as `surface_file` in Table 5.3. Its format is as follows:

Record 1: Grid-type ($1 =$ rectangular, $2 =$ hexagonal) *xmin xmax ymin ymax surface-min surface-max.*

Subsequent records: *x-value y-value surface-value status* where *status* is 0 if (*x-value, y-value*) is inside the boundary, and 1 otherwise.

5.3.8 Discussion

The user need only have a four-page summary of the **id** language to do most ecosystem analyses. No more than five ASCII files are needed to model, estimate, and evaluate an ID or IntIDs model within the **id** software system. The learning curve and required amount of knowledge that is specific to a particular statistical software package is reduced relative to other analysis software systems. A point-and-click front end typical of many GUIs can be used to write an **id** language file. The **id** language file, however, would always be written and saved so as to leave a permanent record of exactly what nodes constitute the ID or IDs, and what kind of reports were requested. The recording of the sequence of GUI operations, some-times called a 'macro recorder,' is not seen as equivalent to an **id** language file due to the former's lack of formal language definition and ad hoc syntax – both of which make a macro recording difficult to interpret and, hence, to share.

The central idea behind the **id** language is to give the user a very high-level language that has a minimum of terms and for which many procedures that characterize best statistical practice have been automated. For further discussion of very high-level analysis and programming languages, see Klerer (1991, chapter 8). Such simplicity and freedom from syntactical minutiae is essential if a larger, more inclusive group of people are to participate in ecosystem management analyses and decision making through the use of this politically realistic EMT.

This language can also be viewed as a pedagogical tool for teaching the use of statistical analysis to nonstatisticians. This is because it organizes the statistical analysis around the inputs and outputs of a particular management situation with-out requiring the user to learn the details of how the statistical computations are performed.

To fully trust such an automated analysis system, work is needed to add enough intelligence to **id** so that the user is warned of questionable output in the report. These warnings would be a result of automatic statistical diagnostics computations.

It is argued that for many statistical analyses performed by nonstatisticians, all that the user really needs to know is the general meaning of each **id** language word and how to submit analysis requests expressed in the **id** language. This philosophy is similar to the attitude towards the teaching of advanced technology in fields other than statistics. For example, a doctor does not need to know the design details of an ultrasound scanner – only what in general are the sensitivities and biases of the scanner when used to form images of human bone and tissue.

It is emphasized that **id** is not an expert system. No attempt is made to program the knowledge of an experienced applied statistician. Therefore, knowledge repre-sentation is not seen as a primary issue in the development of the **id** language or **id** software.

5.4   How the EMT website would be used

The following sequence of activities is the recommended way to use the EMT to manage an ecosystem:

1. Identify an ecosystem with one or more at-risk species of plants or animals.

2. Identify political groups in ecosystem-hosting countries that directly or indi-rectly affect the ecosystem.

3. Construct stochastic decision-making models of each group identified in step 2 and express each of these models in the **id** language.

4. Use social science theory to assign hypothesis values to group ID parameters. Place these values in group ID hypothesis parameter files.

5. Construct a stochastic and dynamic model of the at-risk species identified in step 1 and express this model in the **id** language.

6. Create an **id** language file that defines the EMT's political–ecological system simulator as an IntIDs model by including `influence_diagram` words for all of the above ID models in this **id** language file.

7. Use ecological theory to give hypothesis values to the ecosystem ID's parameters. Place these values in the ecosystem ID's hypothesis parameter file.

8. Store all **id** language files and hypothesis parameter files in the EMT's `input_files` directory.

9. Acquire group actions history data.

10. Acquire ecosystem state data, for example, spatio-temporal species abundance data.

11. Store this political–ecological data in the `data_sets` directory.

12. Use **id** to fit all parameters in the political–ecological system simulator to this data using consistency analysis.

13. Place the results of this analysis in the `out_files` directory.

14. Conduct an analysis of the IntIDs model's reliability (see Chapter 12) by using **id** to perform a sensitivity analysis and compute one-step-ahead prediction error rates. Place the results of these analyses in the `model_reliability` directory.

15. Use **id** to construct the MPEMP (see Chapter 4) for this ecosystem. Place this plan in thc `MPEMP` directory.

16. Implement this MPEMP.

17. As new data becomes available, repeat steps 11 through 16.

Part II

MODEL FORMULATION, ESTIMATION, AND RELIABILITY

6

Influence diagrams of political decision making

6.1 Introduction

A stochastic model of a group's political decision making is described in this chapter. This model, referred to herein as the *group decision-making ID*, was first described by Haas (2004). This model assumes that the decision maker is rational but (a) uses (possibly) inaccurate internal representations of other parties to the decision, (b) must pay decision transaction costs, and (c) is constrained by the institutions through which decisions are funneled. These assumptions are shared by the theory of economic behavior known as *new institutionalism* (see below). The theory of group decision making developed herein is realized as an ID (see below).

This chapter proceeds as follows. First, new institutionalism and the descriptive theory of political decision making are reviewed and shown to have similarities. Then, the model of decision making used to construct group IDs is presented. Following that, some related modeling efforts are reviewed. Conclusions are reached in the final section.

6.2 Theories of political decision making

6.2.1 New institutionalism

The 'new institutionalists' (see Coase 1937, 1960, North 1990) stress that (a) decision makers are pursuing their own personal goals, for example, increasing their

---

Improving Natural Resource Management: Ecological and Political Models   Timothy C. Haas
© 2011 John Wiley & Sons, Ltd

influence and protecting their job; (b) decision makers work to modify institutions to help them achieve these goals; and (c) transaction costs, when accounted for, can explain apparently nonutility maximizing behavior (Hopcroft 1998, p. 279). For discussions of how new institutionalism reinterprets the policy-making process, see Gibson (1999, pp. 9–14, 163, 169–171), Brewer and de Leon (1983), and Lindblom (1980).

The new institutionalism view of the policy-making process is particularly relevant for studying wildlife management in developing countries. As Gibson (1999, pp. 9–10) states:

> New institutionalists provide tools useful to the study of African wildlife policy by placing individuals, their preferences, and institutions at the center of analysis. They begin with the assumption that individuals are rational, self-interested actors who attempt to secure the outcome they most prefer. Yet, as these actors search for gains in a highly uncertain world, their strategic interactions may generate suboptimal outcomes for society as a whole. Thus, rational individuals can take actions that lead to irrational social outcomes.

Individuals are seen as local wealth maximizers that must negotiate positive transaction costs. Hopcroft (1998, p. 279) notes that:

> The main contribution of the new institutional economics to the understanding of economic development is the introduction of the concept of transaction costs and the focus on how social institutions influence economic change in discussions of the issue.

And:

> The new institutional economics suggests that other costs also affect development, including transaction costs, which are socially created costs involved in production and exchange. They include costs of defining and enforcing property rights, measuring the valuable attributes of what is being exchanged, and of monitoring the exchange process itself.

Examples of such transactions and their costs that appear in Chapter 7 include the cost of evicting rural residents from a wildlife reserve, the expected cost of incarceration to a rural resident for executing a poaching transaction, and the bribes paid by a private developer to a permitting agency so that a proposed tourist lodge concession on national park land is approved.

But these actors may have more goals than just wealth maximizing that they are pursuing:

> It should be noted that suggesting that people tend to maximize wealth does not imply that all people in all societies maximize their wealth all the time. Nor does it imply that including wealth as an individual utility excludes the possibility that individuals also have other utilities, such as status, social acceptance, and the like.
>
> (Hopcroft 1998, p. 279).

6.2.2   Descriptive models of political decision making

Another paradigm for political decision making is the *descriptive* model (see Vertzberger 1990). This approach emphasizes that humans can only reach decisions based on their internal, perceived models of other actors in the decision-making situation. These internal models may in fact be inaccurate portrayals of the capabilities and intentions of these other actors. Descriptive decision-making theorists also emphasize the importance of the perceived effect that a decision will have on a decision maker's audiences other than the decision's target, for example, Vertzberger (1990, pp. 249–252).

That this possibly inaccurate internal model can have an effect on economic behavior is acknowledged by new institutionalism theorists. For example, North (1998, p. 249) states that:

> Individuals typically act on incomplete information and with subjectively derived models that are frequently erroneous; the information feedback is typically insufficient to correct these subjective models. Institutions are not necessarily or even usually created to be socially efficient; rather they, or at least the formal rules, are created to serve the interest of those with the bargaining power to create new rules. In a world of zero transaction costs, bargaining strength does not affect the efficiency of outcomes; but in a world of positive transaction costs it does.

6.2.3   Synthesizing rational actors and mental models

To summarize then, new institutionalism models an actor as a wealth maximizer as in classical economics. New institutionalism, however, adds to classical economics theory the assumption that actors are using *mental models* of other actors which may be distorted representations of these other actors. New institutionalism, then, synthesizes economic (rational actor) theories and descriptive theories of decision making.

6.3   Architecture of a group decision-making ID

6.3.1   Overview

The above social science theory suggests that a minimally complete model of political decision making should include representations of the decision maker's goals, internal models of actors, audience effects, a memory mechanism, perceptions of delayed versus immediate effects, and action recognition. Further, the model should have a structure that is plausible based on our current understanding of cognition. Even this minimal model will inevitably be complex. In what follows, a model is developed that is as parsimonious as possible but possesses this minimal set of characteristics.

Modeling is at the group level but the supporting theory is derived from individual decision-making theory. This aggregation, then, ignores interactions between members of a group and assumes that results on how individuals reach decisions can be scaled up to the group level. This aggregation approach to modeling the political groups of a country is similar to the *Political Interaction Framework* of Chazan *et al.* (1999, pp. 23–37) and to a model of interacting social groups developed by Renaud (1989).

A group's ID is partitioned into subsets of connected nodes called the *Situation* and *Scenario* subIDs. The Situation (*St*) subID is the group's internal representation of the state of the decision situation and contains *situation state* nodes. Every Situation subID has a root node that represents discrete time. This node takes on the time point values t_0, t_1, ..., t_T at which model output is desired. The Scenario (*Sc*) subID is the group's internal representation of what the future situation, called the *scenario*, will be like after a proposed option is implemented.

As described in Haas (1992), this structure is based on the Baars (1988) Global Workspace (GW) theory of consciousness. Baars proposes that most mental processing is carried out unconsciously by subprocessors composed of interacting groups of neurons. These subprocessors compete for access to a GW. Although messages on the GW can be read by any subprocessor, subprocessors must compete for the privilege of writing to the GW. Subprocessors organize into large, interconnected, cooperating sets of processors called *contexts*. Contexts that have won control of the GW are said to be *active*. Goals are represented as contexts that contain desired images of perceived future reality (Baars 1988, p. 230). Cognition proceeds by subprocessors reading problems posted by goal contexts to the GW, proposing solutions, and writing these back to the GW. Goal contexts evaluate these solutions and request output contexts (speech, muscle processors) to output the most favored solution. Two other main types of contexts are those that contain images of perceived reality and those that perform specialized problem solving, for example, the 'addition' context.

A context is very similar to the connectionist formulation of a *schema*, discussed in Rumelhart *et al.* (1986). Other than noting that connectionist theory, through a neural network, can produce a quantitative model of a schema, Baars does not advance a specific computational implementation of a context or schema.

Tables 6.1 and 6.2 collect all notation used to express a group ID while Figure 6.1 displays its architecture. To simplify this figure, only the *j*th audience is depicted in the diagram. The group modeled by an instantiation of Figure 6.1 is called the *decision-making group* (DM-group).

Groups interact with each other and the ecosystem by executing actions. The DM-group experiences and/or recognizes an *input action* that is executed by an *actor*. Let *InputAction* denote the DM-group's list of recognized input actions and *Actor* denote the DM-group's list of recognized actors. An input action has a *subject* (which may or may not be the DM-group). Let *Subject* denote the DM-group's list of recognized subjects. The combination of an input action, actor, and subject is an *in-combination*.

The DM-group implements an *output action* directed towards some other group or the ecosystem called a *target*. Let *OutputAction* denote the group's repertoire of actions and *Target* the list of potential targets. The combination of an action

Table 6.1 Definition of symbols used to express nodes in the Situation subID of the group decision-making ID.

Symbol	Definition	Parents
$RE^{(St)}(t_{i-1})$	Economic resources one time point back	Root
$RM^{(St)}(t_{i-1})$	Militaristic resources one time point back	Root
$A_j^{(St)}(t_{i-1})$	Audience j's satisfaction level one time point back	Root
InputAction	Input action	Root
Actor	Actor	Root
Subject	Subject	Root
AF_{Actor}	Actor affect	*Actor*
$AF_{Subject}$	Subject affect	*Subject*
RP_{Actor}	Actor's relative power	*Actor*
$RP_{Subject}$	Subject's relative power	*Subject*
$F_{InputAction}$	Indicator that the input action's effect is delayed	*InputAction*
$M_{InputAction}$	Indicator that the input action uses military force	*InputAction*
$CE_{Subject}$	Subject's economic resources change	*InputAction*, $F_{InputAction}$
$CM_{Subject}$	Subject's militaristic resources change	*InputAction*, $F_{InputAction}$, RP_{Actor}
$CA_j^{(St)}$	Situation change in audience j's satisfaction level	AF_{Actor}, $AF_{Subject}$
$RE^{(St)}$	Situation economic resources level	$RE^{(St)}(t_{i-1})$, *Subject*, $F_{InputAction}$, $CE_{Subject}$
$RM^{(St)}$	Situation militaristic resources level	$RM^{(St)}(t_{i-1})$, *Subject*, $F_{InputAction}$, $CE_{Subject}$
$A_j^{(St)}$	Audience j's satisfaction level	$A_j^{(St)}(t_{i-1})$, $CA_j^{(St)}$
$GE^{(St)}$	Situation economic resources goal	$RE^{(St)}$
$GM^{(St)}$	Situation militaristic resources goal	$RM^{(St)}$
$GMPP^{(St)}$	Situation goal of maintaining political power	$A_j^{(St)}$

and target makes up the group's *decision*. Hereafter, a group's decision will be referred to as either its *out-combination* or its *action–target combination*.

Actions are either verbal (message) or physical events that include *all* inter- and intra-country interactions. Actions are represented by two fundamental

Table 6.2 Definition of symbols used to express nodes in the Scenario subID of the group decision-making ID.

Symbol	Definition	Parents
$OutputAction$	Output action	Root
$Target$	Output action's target	Root
AF_{Target}	Target affect	$Target$
RP_{Target}	Target's relative power	$Target$
$M_{OutputAction}$	Indicator that the action uses military force	$OutputAction$
$CE_{OutputAction}$	DM-group's economic resources change in the scenario	$OutputAction$
$CM_{OutputAction}$	DM-group's militaristic resources change in the scenario	$OutputAction$, $M_{OutputAction}$, RP_{Target}
$RE^{(Sc)}$	Scenario economic resources	$RE^{(St)}$, $CE_{OutputAction}$
$RM^{(Sc)}$	Scenario militaristic resources	$RM^{(St)}$, $CM_{OutputAction}$
$CA_j^{(Sc)}$	Scenario change in audience j's satisfaction level	$OutputAction$, AF_{Target}
$A_j^{(Sc)}$	Audience j's scenario satisfaction level	$CA_j^{(Sc)}$
$GE^{(Sc)}$	Scenario economic resources goal	$RE^{(Sc)}$, $GE^{(St)}$
$GM^{(Sc)}$	Scenario militaristic resources goal	$RM^{(Sc)}$, $GM^{(St)}$
$GMPP^{(Sc)}$	Scenario goal of maintaining political power	$A_j^{(St)}$, $A_j^{(Sc)}$
OGA	Overall goal attainment	$GE^{(Sc)}$, $GM^{(Sc)}$, $GMPP^{(Sc)}$

characteristics: the actor's resource amount change as a result of the action and the subject's resource amount change as a result of the action. These characteristics are described below.

Conditional on an input action, the Situation subID is one way to represent a schema (see Rumelhart *et al.* 1986) or story that the decision maker invokes upon receipt of an in-combination.

6.3.2   Ecosystem status perception nodes

Quantities that represent ecosystem status are input nodes to a group ID. These nodes influence nodes that represent how sensitive the group is to values of corresponding ecosystem status nodes. The idea is that a group is affected by the ecosystem but is only conscious of it through filtered and possibly distorted perceptual functions of the underlying ecosystem status nodes. For example, a group ID is sensitive to the presence of a land animal such as the cheetah through

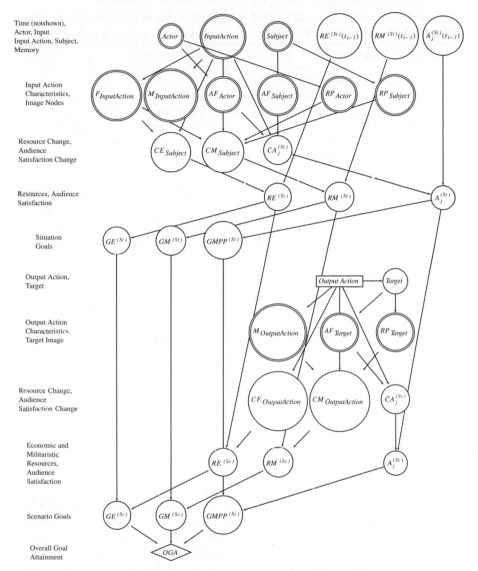

Figure 6.1   Architecture of the group decision-making ID. Ecosystem input and ecosystem status perception nodes are not shown.

the animal's density (number per hectare). This sensitivity is modeled by having the animal's density node influence a perceived animal prevalence node that takes on the values *none*, *few*, and *many*.

An ecosystem status perception node is stochastic so that it can represent random distortion and uncertainty in the transfer of ecosystem information from the physical world to a group's belief system. The need to model this mapping between the actual and perceived state of an ecosystem was first discussed in Haas (2004):

> For example, consider the size of a minority population in a city. A demographic model allows for the stochasticity in birth rates, death rates, migration, and emigration. Now consider a hypothetical elderly member of this city who's only source of information is TV news and newsletters from politically conservative groups. The perceived size of the minority population of the city by this elderly person may be only minimally affected by the probability distribution of this quantity computed from the demographic model. Further, this elderly person may reason about the size of this minority population in categorical terms (small, moderate, hordes). The credence this individual will give to these different values will be determined in-part by random encounters with members of this minority group (captured by the demographic model) and by some noisy function of images viewed on TV and statements made in newsletters ('... members of minority group X are over-running this city!').

A separate node within each group ID is then seen as necessary to capture both the coarser resolution of perceptual models of continuously valued quantities and the unique sources of stochasticity characteristic of perceptual processing of the physical world.

This mechanism is captured in the decision-making model as follows. First, ecosystem ID output nodes (see Chapter 8) are root nodes in the Situation subID. These nodes influence ecosystem status perception nodes that are also in the Situation subID. These nodes, in turn, influence economic resource change nodes in the Situation subID.

6.3.3   Image nodes

The DM-group defines its image of another group along the dimensions of Affect and Relative Power. Affect varies over the *enemy–neutral–ally–self* dimension (see Murray and Cowden 1999, Hudson 1983, Chapters 2–4). Relative Power varies over the *weaker–parity–stronger* dimension. The Affect dimension's *self* category is needed because the subject of an input action may be the DM-group itself.

Let AF_{Actor} and RP_{Actor} be the DM-group's Affect and Relative Power image nodes of the Actor group, respectively. Define $AF_{Subject}$ and $RP_{Subject}$ similarly for the input action's subject.

6.3.4   Economic, militaristic, and institutional goal nodes

It is assumed here that a group evaluates an in-combination by assessing its perceived immediate and future impacts on economic, militaristic, and institutional goals both in the present (the Situation) and, depending on the chosen out-combination, in the future (the Scenario).

For economic and militaristic goals, this is a two-step process: first, the DM-group assesses how the in-combination changes its amount of economic or militaristic resources; then, an assessment is made of how this new resource level affects the associated economic or militaristic goal. Proposed actions affect Scenario goals in the Scenario subID similarly.

Only one institutional goal is modeled: Maintain Political Power. This goal is solely dependent on the perceived level of satisfaction of several important *audiences* (see below). Denote Situation subID economic, militaristic, and institutional (maintaining political power) goal nodes by $GE^{(St)}$, $GM^{(St)}$, and $GMPP^{(St)}$, respectively. Define $GE^{(Sc)}$, $GM^{(Sc)}$, and $GMPP^{(Sc)}$ similarly for the Scenario subID. A goal node takes on the values *unattained* and *attained* as per common usage (see, e.g., Uleman 1996).

Actor image nodes, Subject image nodes, and the nodes representing the input action's immediate and future impact on the DM-group's resources affect goal status assessment. Scenario goals are influenced by Situation goals: if a proposed action does not address the goal, the mode of that Scenario goal node equals the conditioning value of the corresponding Situation goal node.

One example of how an input action affects goal nodes is the president of Kenya creating a new wildlife reserve. For the rural resident DM-group, this action would cause $GE^{(St)} = GE^{(Sc)} = $ *unattained*, and $GM^{(St)} = GM^{(Sc)} = $ *attained*, to represent the rural resident perceiving denial of access to forest resources in the future through a nonmilitaristic action by Kenya's president.

6.3.5   Audience effects

The influence of audiences on a decision maker is supported by research that suggests perceptions of present and future reactions of important audiences have an effect on decision making and bargaining, see Fearon (1994), Keller (2005), Leeds (1999), and Partell and Palmer (1999). The perceived impact of an in-combination on an audience is the decision maker's assessment of how the in-combination changes an audience's judgment of the decision maker's success in satisfying their demands. This perceived level of the satisfaction of audience demands is referred to here as an audience's *satisfaction level*. For example, an important audience for President Moi during his presidency of Kenya was his ethnic group, the Kalenjin (Throup and Hornsby 1998, p. 8). President Moi knew that only actions that brought benefits to that tribe would satisfy their demands on him.

The effects of perceived audience reactions to input actions are modeled by having input action, actor, and subject characteristics influence Audience Satisfaction Change nodes which, in turn, influence Audience Satisfaction nodes.

Say that the DM-group ID has m important audiences. Let the node $A_j^{(St)}$ denote the satisfaction level of audience j. $A_j^{(St)}$ takes on the values *dissatisfied*, *ambivalent*, and *satisfied*. Let the node $CA_j^{(St)}$ denote the change in satisfaction of audience j due to the input action and the DM-group's Affect perceptions of the actor and subject (perceptions that the DM-group assumes are shared by all audiences). $CA_j^{(St)}$ takes on the values *decreased*, *no change*, and *increased*. Let $\mathbf{CA}^{(St)} = (CA_1^{(St)}, \ldots, CA_m^{(St)})'$ and $\mathbf{A}^{(St)} = (A_1^{(St)}, \ldots, A_m^{(St)})'$.

Likewise, in the Scenario subID, action characteristics influence Scenario Audience Satisfaction nodes through the Scenario Audience Satisfaction Change nodes. Let $\mathbf{CA}^{(Sc)} = (CA_1^{(Sc)}, \ldots, CA_m^{(Sc)})'$. These nodes are influenced by the OutputAction and Target Affect nodes. This set of parents allows the modeling of perceived audience expectations for the DM-group to do something as a result of an input action. Audience Satisfaction nodes are not influenced by economic or militaristic change nodes.

The Situation Audience Satisfaction nodes influence Scenario Audience Satisfaction nodes as follows. If a proposed action does not change Scenario audience satisfaction, the mode of the corresponding Scenario Audience Satisfaction node is the conditioning value of the corresponding Situation Audience Satisfaction node.

For IDs of presidents only, in either subID, Audience Satisfaction nodes only influence the Maintain Political Power goal – the decision maker does not explicitly have audience satisfaction as a goal. This is because the decision maker has no concern for these audiences other than how they affect the decision maker's hold on political power.

Each audience has its own Audience Satisfaction Change node in order to represent the effect of some new action on the previous level of that audience's satisfaction (represented in the Situation subnetwork by an Audience Memory node, and in the Scenario subID by the Situation Audience Satisfaction node). Absorbing Change nodes into Satisfaction nodes would confound the distinction between a present level of satisfaction and a change in satisfaction due to an action.

6.3.5.1   Political corruption effects on decision making

Audience effects on a decision maker can explain some aspects of governmental corruption. One of the many forms that corruption takes is the increased influence of a particular audience on the decision maker through payments of various kinds. Here, such corruption is modeled implicitly by having certain audiences exert a strong influence on the decision maker. The modeling of other forms of corruption is a topic for future research.

6.3.6   Resource nodes

A *resource* is anything of economic or militaristic value to a group. Let $RE^{(St)}$ and $RM^{(St)}$ be the DM-group's perception of its present absolute amount of economic and militaristic resources, respectively.

Define $RE^{(Sc)}$ and $RM^{(Sc)}$ similarly for the Scenario subID. Resource nodes are modeled as chance nodes because they represent perceived resource amounts and hence are not typically known by the DM-group to an exact value. Typically, the decision maker has only a qualitative idea of their present levels of economic or militaristic resources. These nodes therefore take on only three values: *negligible*, *inadequate*, and *adequate*. A militaristic resource is broadly defined to include military material and territory won through military conquests.

Define $CE_{Subject}$ to be the DM-group's perception of the subject's relative change in economic resources due to the input action. Likewise for the Scenario subID, define $CE_{OutputAction}$ and $CM_{OutputAction}$ as the change in the DM-group's economic and militaristic resources, respectively due to a proposed action. All of these Change nodes can take on the five ordinal values of *large_decrease*, *small_decrease*, *no_change*, *small_increase*, and *large_increase*.

Resource nodes specify the change in resources on separate economic and militaristic dimensions because an action can cause resource change on either or both dimensions. For example, a military blockade can have significant economic consequences for the subject.

Let $F_{InputAction} = 1$ if the input action's resource changes will occur in the future, otherwise let this node's value be 0. Define $M_{InputAction}$ to be 1 if the input action involves the use of military force and 0 otherwise. Define $M_{OutputAction}$ similarly.

For each action, as a component either of an in-combination or of an action–target combination, relative resource change node values are assigned subjectively. Except for certain extreme actions, the relative change caused by an action on the subject or DM-group is dependent on present levels of the group's resources and the specific nature of the action. Most EMAT actions (see Chapter 9) are not specific enough to allow a Change value to be assigned even if the action's actor and subject were specified. Therefore, Change values are assigned locally, that is, a list of action–actor–subject triads is developed for the particular actions history being modeled. Then, Change values for each action are assigned to be the maximum effect that the action would have on the subject or DM-group. See Chapter 7 for examples of such assignments.

6.3.7   Group memory

A group's memory through time is modeled by having the nodes $RE^{(St)}$, $RM^{(St)}$, and $\mathbf{A}^{(St)}$ at the previous time step influence these same nodes at the current time step. See Haas *et al.* (1994) for an example and the mathematical form of this approach to representing the passage of time with IDs.

This memory mechanism allows the modeling of changes in group perceptions through time. Examples include perceived resource depletion and accumulated resentment or desperation. These nodes are given distributions at time t_0 that are starting values of the group's perceived level of economic and militaristic resources – and level of satisfaction of each audience in $\mathbf{A}^{(St)}$.

6.3.8   Action and target nodes

Many actions in the EMAT classification system (see Chapter 9) strongly imply only one or at most two possible target types, for example, 'arrest poachers' would not be directed towards a president or a conservation NGO. Hence, in the Scenario subID, the `OutputAction` node influences the discrete chance node, `Chosen Target`.

6.3.8.1   Perceived effectiveness of out-combinations

The DM-group's perception of the economic or militaristic effectiveness of a proposed action is modeled by having the `OutputAction` and `Target` nodes influence `Scenario Economic Resources Change` and `Scenario Militaristic Resources Change` nodes, respectively.

6.3.9   Overall goal attainment node

Goal prioritization is modeled by a single node representing the DM-group's overall sense of well-being through goal attainment. This node, denoted by OGA, is a deterministic function of the goal nodes wherein the coefficients in this function are interpreted as goal-importance weights and hence are assigned directly from knowledge of the group's goal priorities.

6.3.10   How an ID simulates a decision

A proposed action–target combination influences action characteristic and target image nodes. These nodes, along with Situation goal nodes, influence Scenario goal nodes. Finally, Scenario goal nodes influence the OGA node (see Figure 6.1). Each action–target combination is used to compute the expected value of the OGA node. At time t, the action–target combination that maximizes this expected value is designated by $c_{optimal}(t)$. Computing the maximum utility output value is called *evaluating* the ID, see Nilsson and Lauritzen (2000).

6.4   Related modeling efforts

6.4.1   Welfare disbursement decision making

Renaud (1989) gives a political economics-based model of how governments determine disbursement amounts to welfare agencies. He makes two points that are relevant to the decision-making model developed in this book:

1. The decision making of a 'representative' individual from each group is modeled rather than attempting to model the group dynamic within each group.

This is the route taken with this book's models of presidents, EPAs, rural residents, pastoralists, and a group of conservation NGOs.

2. The policy option that maximizes the weighted sum of all social group utility functions wherein the weights are governmental 'pressure' or 'influence' measures is the policy option that the government adopts.

The option-selection algorithm employed in this book's group ID model becomes identical to that used by Renaud (1989) through the use of the *OGA* node.

Renaud (1989) does not need to determine the values of each policy along relevant decision-making dimensions as here because he is modeling only policies that are completely described by the monetary value they assign to each group. He does need, however, to assign the relative importance weights (pressure values) that each group has on the government's decision making.

Renaud (1989) uses the following symbols to define his model:

g: a government sector employee,

p: a private sector employee,

s: a capital owner or self-employed worker,

d_1: an unemployed worker,

d_2: a pensioner,

d_3: an otherwise dependent, for example, a widower,

K: the set $\{g, p, s, d_1, d_2, d_3\}$,

K_1: the set $\{g, p, s\}$,

G: the number of units of government-provided goods,

g: the unit price of a government-provided good,

τ: the tax rate, and

w_k: the amount of wages or benefits payments to an individual in the kth group.

Let the elementary interest (utility) function of the kth group be $P_k^e = v_k \ln G + \beta_k \ln[(1 - \tau)w_k]$ where v_k and β_k are weights of the relative importance that members of this group attach to government-provided goods and payments, respectively. Note that v_k, $\beta_k \geq 0$ and $v_k + \beta_k = 1$.

A group member may expect to enter another group (e.g., a worker who is anticipating a lay-off) or already be in more than one group (e.g., a worker who is only employed part-time). To accommodate possible multigroup membership and an individual's resultant interest in more than one group's interest function, let π_{kl} be this interest that an individual in the kth group has in the utility of the lth group. Let $\sum_l \pi_{kl} = 1$ and $\pi_{kl} \geq 0$. Then, the 'complex' interest function for the kth group is $P_k = \sum_l \pi_{kl} P_l^e$.

Policy instruments available to the government are:

G: (see above),

$(1 - \tau)w_{d_1}$: net unemployment benefit,

$(1 - \tau)w_{d_2}$: net pension payment, and

$(1 - \tau)w_{d_3}$: net transfer payment to 'otherwise' dependent individuals.

To describe the government's budget, let E_k be the number of individuals in group k, and let S be the sum of nontax government income and the budget deficit. S is exogenous to the model. Also, let W_1 be the total labor and profit income of the kth group:

$$W_1 = \sum_{k \in K_1} w_k E_k. \tag{6.1}$$

The government's budget must satisfy

$$\tau W_1 + \sum_{i=1}^{3} \tau w_{d_i} E_{d_i} + S = G + \sum_{i=1}^{3} w_{d_i} E_{d_i}. \tag{6.2}$$

Note that the second term on the left-hand side is tax revenue from unemployment, pension, and transfer payment benefits.

The solution to this bargaining game is as follows. Let α_k be the relative influence that the kth group has on the decision process ($0 \le \alpha_k \le 1$, $\sum \alpha_k = 1$). Let W be the 'fiscal capacity' of the system, that is, the amount of money that is available for allocation through the political decision-making process:

$$W = (1 - \tau)W_1 + \sum_{i=1}^{3}(1 - \tau)w_{d_i} E_{d_i} + p_G G$$
$$= W_1 + S. \tag{6.3}$$

Given fixed values for ν, β, π, G, E, α, and S, Nash equilibrium is achieved when the weighted sum of all interest groups, $P = \sum_{k \in K} \alpha_k P_k$, is maximized over the space of transfer payments and interest rate. Renaud (1989, pp. 62–63) gives an analytical solution to this maximization problem.

6.4.2   Landuse decision making

The Center for the Study of Institutions, Population, and Environmental Change at Indiana University Bloomington (funded by the United States National Science Foundation) studied private landowner decision making in Indiana and Brazil (see Hoffmann et al. (2002)). These researchers developed multiagent models of the effect of agricultural practices (including irrigation) on a landscape. To do this, the authors wrote a multiagent simulation model of forest harvesting decisions of landowners in southern Indiana and in the Brazilian Amazon. A differential equation model of a rangeland's reaction to these decisions was also derived.

Landowner models represent the passage of time with finite differences, employ a simple learning algorithm, and use a maximum expected utility algorithm to reach landuse decisions.

For example, a decision to harvest trees is made if current economic conditions cause the utility of cutting to be larger than that of not cutting. This utility function is

$$U_{cut} = (1 - \text{wealth})[\text{forest-age} \times \text{forest-price} - \text{cost-cut}]$$
$$- \sigma_{forest}^2 - (\text{wealth} \times \text{education} \times \text{forest-age}).$$

The first term represents the diminishing returns of profit, the second term represents the shock due to uncertainty, and the third term represents the landowner's aesthetic pleasure of owning an old growth forest.

6.5   Conclusions

New institutionalism has been reviewed and shown to be a platform for deriving models of political decision making by groups that synthesizes classical economics' ideas of wealth maximization and cognitive psychological ideas of how groups perceive other groups and anticipate the actions of these other groups.

An architecture for constructing a model of a group's political decision making based on an ID has been presented. Such an architecture yields a model that is easy to understand through its graphical representation, is inherently stochastic, and can capture nonlinear relationships between its subcomponents and between its inputs and its outputs.

As the last section of this chapter shows, the idea of building computational models of political decision making is not without precedent.

6.6   Exercises

1. Construct a group decision-making ID of an unemployed worker's decision making using the context of Renaud (1989).

2. Construct a group decision-making ID of a landowner's decision to cut down or not cut down forest that is growing on his/her land using the context of Hoffmann et al. (2002).

7

Group IDs for the East African cheetah EMT

7.1 Introduction

In this chapter, IDs are constructed for the groups to be simulated in the East African cheetah EMT. These groups are the president's office, EPA, rural residents, and pastoralists in each of the countries of Kenya, Tanzania, and Uganda. A 13th group is modeled to represent the collective actions of conservation NGOs that are involved in these three countries.

To give context to the descriptions of these group IDs, overviews of these three countries are given first. Then, reasons are given for why the particular groups were chosen for simulation. Following that, descriptions are given of each group's ID. Conclusions are reached in the final section.

7.2 Country backgrounds

7.2.1 Kenya overview

Kenya is a republic composed of a presidential office (currently President Kibaki), a legislature (the National Assembly or Bunge), and a judicial branch (the Court of Appeals and the High Court). Until recently, Kenya was one of the more stable regimes in Africa. President Kibaki replaced President Moi in 2002 but in 2007 a tainted election led to widespread rioting. Based on the World Bank's Control

Improving Natural Resource Management: Ecological and Political Models   Timothy C. Haas
© 2011 John Wiley & Sons, Ltd

of Corruption Index, Kenya is highly corrupt since its score for 2007 of $-.97$ (range: -2.5 to 2.5) places it in the 15.5 percentile of the 212 countries scored (World Bank 2009).

As mentioned in Chapter 1, approximately 50% of Kenya's population is in poverty. About 6.7% of Kenyan adults have contracted HIV/AIDS (CIA 2010a). There are five main ethnic groups: the Kikuyu, Luhya, Luo, Kalenjin, and the Kamba. Ethnic tensions are high (see Baldauf 2008, Oucho 2002). Kenya's unemployment rate is about 40% and 75% of those who are employed, work in agriculture. Kenya's per-capita income is $1700, which is less than $11 455, the cut-off for 'developed country' status in the eyes of the World Bank (World Bank 2008).

Kenya's economy is mostly agricultural. The country's main exports are tea, horticultural products, coffee, petroleum, fish, and cement. Kenya has a GDP (purchasing power parity) of about $60 billion and receives about $1 billion per year in aid. Kenya's inflation rate is at least 9%.

Kenya has a long coastline with the Indian Ocean, a central highland that is agriculturally very productive, a fertile western plateau, and a very dry, very large northern region called the Marsabit. Nairobi and Mombasa are the two largest cities (ExperienceKenya 2010). Pastoralists exist in the arid and semiarid northern and eastern regions of Kenya (FAO 2008).

7.2.2   Tanzania overview

Tanzania consists of a portion of the African continent and the relatively small island of Zanzibar. The mainland portion is a republic composed of a presidential office (currently President Kikwete), a legislature (the National Assembly or Bunge), and a judicial branch (a Permanent Commission of Enquiry, Court of Appeal, the High Court, District Courts, and Primary Courts). Zanzibar has a House of Representatives for internal legislation but is only nominally democratic. Tanzania is corrupt with a Control of Corruption Index value of $-.45$ (43.0 percentile) (World Bank 2009).

As mentioned in Chapter 1, 36% of Tanzania's population is in poverty. About 8.8% of Tanzanian adults have contracted HIV/AIDS (CIA 2010b). The great majority of Tanzanians belong to the Bantu ethnic group. Tanzania's unemployment rate is about 12% (EISA 2010). About 80% of those who are employed work in agriculture. Tanzania's per-capita income is $1300 and it is considered to be one of the poorest countries in the world.

The country's main exports are gold, coffee, cashew nuts, manufactured goods, and cotton. Tanzania has a GDP (purchasing power parity) of about $49 billion and receives about $1.5 billion per year in aid. Tanzania's inflation rate is about 7%.

Tanzania also has a long coastline with the Indian Ocean, a central plateau, plains along the coast, and highlands in the north and in the south of the country.

Dar es Salaam and Mwanza are the two largest cities (Tageo 2010a). Pastoralists exist in a small region in northern Tanzania around the Serengeti plain.

7.2.3 Uganda overview

Uganda is a republic composed of a presidential office (currently President Museveni), a legislature (the National Assembly), and a judicial branch (the Court of Appeals and the High Court). Uganda experienced horrific purges just after its independence in 1962. President Museveni, after seizing power in 1986, has overseen a stabilization of the country. Uganda is highly corrupt with a Control of Corruption Index score of −.76 (24.6 percentile) (World Bank 2009).

As mentioned in Chapter 1, 40% of Uganda's population is in poverty. About 4.1% of Ugandan adults have contracted HIV/AIDS (CIA 2010c). There are five main ethnic groups: the Baganda, Banyakole, Basoga, Bakiga, and the Iteso. Ethnic tensions are high (see Green 2005, Espeland 2007). Uganda's unemployment rate is about 5% (Nsimbe 2008). About 80% of those who are employed work in agriculture. Uganda's per-capita income is $900 making it a very poor developing country.

Uganda's economy is mostly agricultural. The country's main export by far is coffee but it also exports petroleum products, fish, tea, cotton, flowers, horticultural products, and gold. Uganda has a GDP (purchasing power parity) of about $29 billion and receives about $1.2 billion per year in aid. Uganda's inflation rate is about 6.8%.

Uganda has shoreline with Lake Victoria and is mostly a plateau surrounded by mountains. Kampala and Gulu are the two largest cities (Tageo 2010b). Pastoralists exist in a small region in the extreme northeast corner of Uganda.

7.3   Selection of groups to model

According to Gros (1998) and Gibson (1999, p. 164), the groups that directly affect the cheetah population are Environmental Protection Agencies (EPAs), ranchers, rural residents, and pastoralists. Conservation NGOs can be added to this list as they can engage in animal translocation. Each country's presidential office (hereafter *president*), legislature, and courts indirectly affect the cheetah population through their influence on these direct-effect groups. The political–ecological system simulator represents (a) the presidents, EPAs, rural residents, and pastoralists of Kenya, Tanzania, and Uganda, (b) a model of the group of conservation NGOs that are working on projects in all three countries, and (c) the cheetah-supporting ecosystem enclosed by the political boundaries of these three countries (see Chapter 8). This version of the model omits group IDs for legislatures, courts, and large commercial ranches.

The following sections describe each group ID. As mentioned in Chapter 4, prior to statistically fitting these models to data, parameter values that are based solely on what is expected from theoretical considerations (hypothesis values) are assigned to each parameter.

Note that the ecosystem ID is directly affected only by poaching activities, animal relocation, rural resident and/or pastoralist eviction, and land clearing (see Chapter 8). Anti-poaching enforcement is directed towards rural residents and/or pastoralists – and may or may not be effective at reducing poaching activity.

See Appendix A for the heuristics used to represent subject matter theory for the purposes of hypothesis parameter value assignment.

7.4   President IDs

Here, the development of a decision-making model of an East African country's president begins with the perusal of informal commentary on the nature of these presidents. Gibson (1999, pp. 155–156) argues in his case studies of Kenya, Zambia, and Zimbabwe that the president in each of these countries has a different personal priority for protecting ecosystems. Presidents of politically unstable countries typically place a high priority on protecting their power and staying in office (Gibson 1999, p. 7). There is a tendency in African politics towards *neopaternalism* wherein the president is viewed as a strong man dispensing favors to loyal, children-like supporters. This is particularly true of President Yoweri Museveni of Uganda (see Kassimir 1998).

These perspectives have motivated the following form of the president IDs (see Figure 7.1).

In this ID, the president has direct knowledge of the actions of the country's rural residents and pastoralists. The president's sole audience are the group of donors to his presidential campaign. Neither aid-granting countries nor the military are included as audiences in this version. The current versions of these president IDs also exclude the president's affect towards the actor or target. The president's goals are to maintain political power and domestic order. Defending the country is not included as a goal in this version.

Actions are derived from the political actions data set for the East African cheetah EMT (see Chapter 9). Table 7.1 collects all in-combinations recognized by a typical president ID. This table also gives hypothesis values of resource change nodes under each input action, and each input action's hypothesis values for whether the input action's effect will be immediate ($F_{InputAction}$) and whether the input action involves the use of force ($M_{InputAction}$).

Hypothesis parameter values for other nodes in a president ID are on the East African cheetah EMT website (www4.uwm.edu/people/haas/ cheetah_emt).

Table 7.2 lists the repertoire of action–target combinations for a typical president ID.

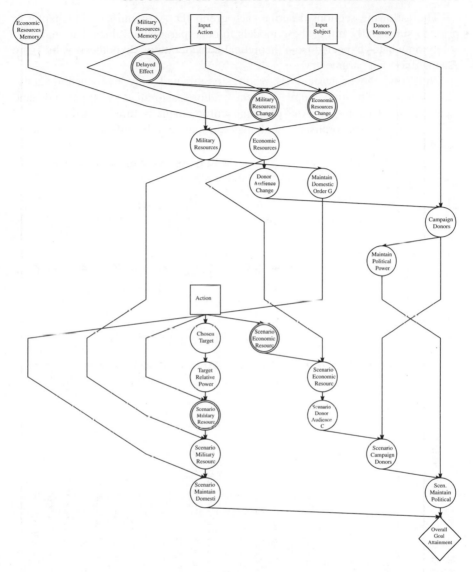

Figure 7.1   Kenya president ID.

7.5   EPA IDs

EPA perceptions of the ecosystem's state are represented by herbivore prevalence and cheetah prevalence nodes, see Figure 7.2. These nodes are influenced
by the expected values of the ecosystem ID's Herbivore Fraction
Detected node and the Cheetah Fraction Detected node, respectively. Specifically, the Herbivore Prevalence node takes on the values

Table 7.1 Input actions that affect economic and/or militaristic resource nodes in a president ID. S, *small*; L, *large*; +, *increase*; and −, *decrease*. RR: rural residents, Pas: pastoralists.

Actor	Input action	$CE_{Subject}$	$CM_{Subject}$	$M_{InputAction}$	$F_{InputAction}$
RR, Pas	Poach for food	−L	N	1	0
RR, Pas	Poach for cash	−S	N	1	0
RR, Pas	Poach for protection	−L	N	1	0
RR	Riot	N	−L	1	0
RR	Clear new land	−S	N	0	0
RR	Abandon settlement	−S	N	0	0
RR	Devastate a region	−L	N	0	0
RR	Murder game wardens	−S	−L	1	0
RR	Report: wildlife attack RRs	N	−S	0	0
Pas	Agree to create wildlife Wildlife reserves	S	N	0	1
EPA	Decrease anti-poaching enforcement	N	−S	1	1
EPA	Increase anti-poaching enforcement	N	−S	0	1
EPA	Negative eco-report	−S	N	0	1
EPA	Positive eco-report	S	N	0	1
EPA	Suspend corrupt officers	N	−S	1	0
EPA	Plan water storage upgrade	−S	N	0	1
EPA	Seize elephant ivory	S	S	1	0
EPA	Detain RRs for encroachment	N	S	1	0
EPA	Translocate animals	−S	N	1	0
EPA	Use technology to locate habitat	−S	N	0	0
EPA	Host conservation conference	−S	N	0	0
EPA	Kill marauding wildlife	−S	N	1	0

Table 7.2   Actions and targets for a typical president ID. See Table 7.1 for abbreviations.

Output action	Likely targets
Open a wildlife reserve for settlement	RR, Pas
Create wildlife reserve	RR, Pas
Tighten wildlife agreement or laws	RR, Pas
Suppress riot	RR
Seize idle land for poor	RR
Declare tree planting day	RR
Fund conservation project	RR
Invest in tourism infrastructure	RR
Punish or restrict domestic ministers	RR
Donate to establish wildlife trust fund	EPA
Request increased anti-poaching enforcement	EPA
Request ivory trade ban continuation	NGOs
Host or attend conservation conference	NGOs
Sign inter-country customs pact	Presidents

none, *few*, and *many* and has a cumulative logit distribution with one explanatory variable (see Agresti 2002, chapter 7). Here, this explanatory variable is E[Herbivore Fraction Detected]. Values of this variable, having been previously posted to the bulletin board (see Chapter 2) by the ecosystem ID, are read from the bulletin board each time an EPA ID is solved. Likewise, Cheetah Prevalence is a cumulative logit chance node wherein the single explanatory variable in the logit model is the ecosystem ID's computed value of E[Cheetah Fraction Detected].

The EPA's sole audience is the president. The goals of the EPA are to protect the environment and to increase the agency's staff and budget. The latter goal is motivated by an examination of the literature on bureaucracies. For example, Healy and Ascher (1995) note that during the 1970s and 1980s the USDA Forest Service, using FORPLAN output, consistently proposed forest management plans that required large increases in Forest Service budget and staff (see also Gibson 1999, pp. 85, 115–116).

7.6   Rural residents IDs

A rural residents ID has prevalence nodes for both herbivores and cheetah, see the EPA ID above. Rural residents are pursuing the two goals of supporting their families and avoiding prosecution for poaching, see Figure 7.3.

A rural resident's sole audience is his/her family. Note that here, in contrast to a political leader's ID, rural residents do have audience satisfaction as one of their goals.

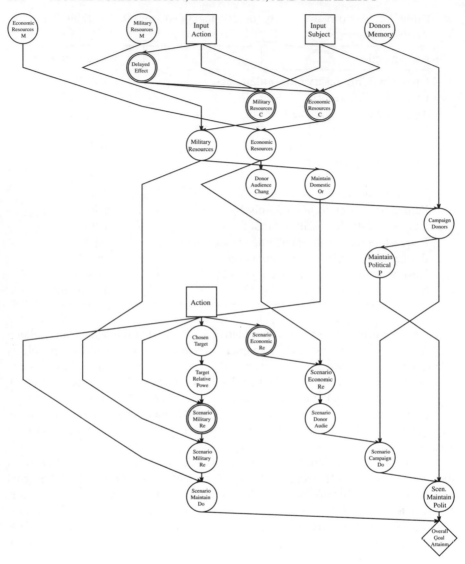

Figure 7.2   Kenya EPA ID.

7.7   Pastoralists IDs

Similar to the rural residents ID, the pastoralists ID has prevalence nodes for both herbivores and cheetahs, see the EPA ID above. Pastoralists have the three goals of supporting their family, protecting their livestock, and avoiding prosecution for poaching, see Figure 7.4.

A pastoralist's sole audience is his/her family.

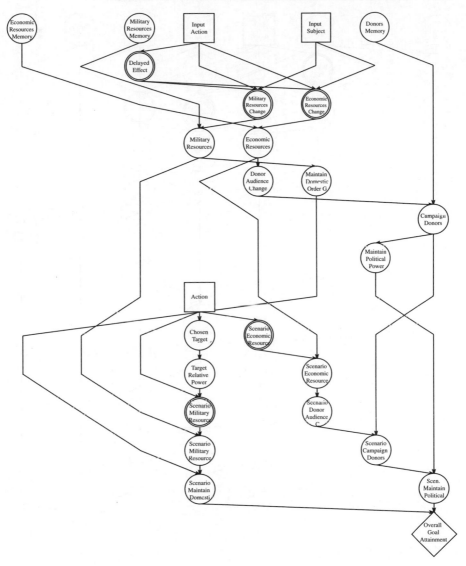

Figure 7.3   Kenya rural resident ID.

7.8   Conservation NGOs ID

Because this is a model of a group of NGOs, an input action by a group in a particular country is assumed to be reacted to by an NGO in that same country.

NGOs within this group keep track of input actions that affect wildlife in each country. These changes affect the overall perceptions of cheetah and herbivore prevalence over the entire three-country area.

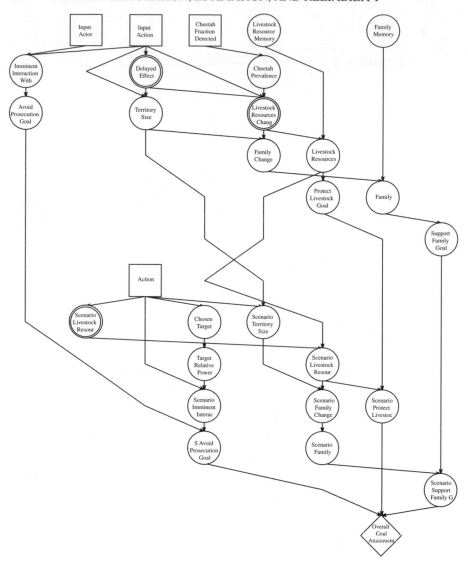

Figure 7.4   Kenya pastoralist ID.

The audiences of NGOs are financial backers (donors) who reside in developed countries outside Africa, and the governments of the three host countries as embodied in each country's office of the president. See Figure 7.5.

NGOs have three goals: (1) conserve wildlife, (2) maintain productive relations with each host country's government, and (3) raise operations funds.

Duffy (2000, p. 117) discusses how important the maintenance of productive relations with the host country is to a wildlife conservation NGO.

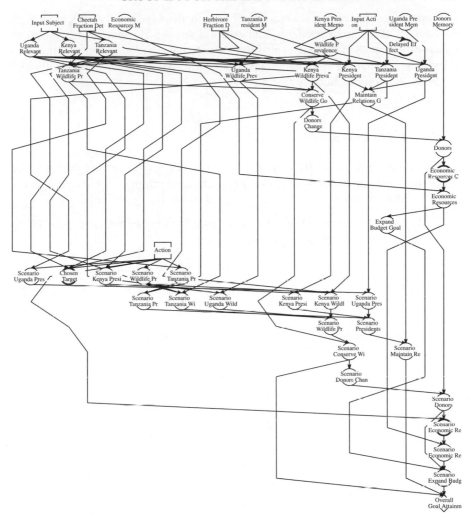

Figure 7.5   Conservation NGOs ID.

Because an NGO's sole support is from external funds, input actions do not affect its budget. Rather, NGOs perceive that only the satisfaction level of external donors and the previous time step's economic resources affect its budget status.

7.9   Conclusions

Basic facts were given about the developing countries of Kenya, Tanzania, and Uganda. These countries form the geographic region that is the focus of the East African cheetah EMT. Justification was then provided for the choice of groups to model. Generic versions were described and displayed of the ID models of presidents, EPAs, rural residents, and pastoralists. An ID model of a group of conservation-focused NGOs was also given. Each of these models captures the most relevant goals that the modeled group is pursuing. All of these models, although complex, are computationally feasible.

7.10   Exercises

1. Construct a group ID of private landowners in Kenya that includes the possible output action of converting pasture for commercial livestock into a wildlife reserve.

2. Construct a group ID of a for-profit tourist lodge in the Serengeti that is owned by an investment group in Europe.

8

Modeling wildlife population dynamics with an influence diagram

8.1 Introduction

A set of regions is given in this chapter that partitions Kenya into areas that have relatively homogeneous climate and vegetation regimes. Regions for Tanzania and Uganda are also given that, instead, follow the political district partitioning of these two countries. Then, an ecosystem ID is described that models cheetah and prey abundance within each of these regions. This ID contains a system of SDEs that model cheetah–prey population dynamics. This population dynamics model is affected by regional characteristics and local management actions.

The solution of a differential equation is a function that maps values of the independent variable to values of the dependent variable. The solution of an SDE is a cumulative distribution function of the dependent variable as a function of the independent variable. The system of SDEs in this application, however, does not have an analytical solution. Therefore, MC methods are used to approximate these cumulative distribution functions.

This chapter's layout is as follows. An architecture for a wildlife abundance model (nodes and node connectivity) is given in Section 8.2 and is made specific by using it to construct an ID for the East African cheetah. The population dynamics of cheetahs and their prey is expressed as a system of SDEs (or simply

Improving Natural Resource Management: Ecological and Political Models   Timothy C. Haas
© 2011 John Wiley & Sons, Ltd

SDE system) embedded in the ID. Time, region, and management action are the independent variables of this SDE system.

Section 8.3 contains a description of one such MC method that has been modified to allow the SDE system to be embedded in an ID. Section 8.4 contains examples of finding such approximate solutions to the East African cheetah ecosystem ID. Conclusions are reached in Section 8.5.

8.2   Model of cheetah and prey population dynamics

The ecosystem ID is an extended and modified version of the cheetah population dynamics ID of Haas (2001) and consists of five subIDs: inputs, habitat, direct effects on population dynamics, population dynamics, and observable random variables (see Figure 8.1).

8.2.1   Inputs

The input nodes represent time (t), region (q), and management action (m). Descriptions of these nodes follow.

As with the Time node in each group ID (see Chapter 6), the ecosystem ID's Time node holds the end-time value in units of years.

Regions partition each country's area into smaller areas that have approximately homogeneous climate and vegetation regimes. For Kenya, such a partition (11 regions) was digitized from a partitioning given in Gros (1998) (Figure 8.2). For Tanzania and Uganda, administrative districts are used for this partitioning because of their small areal size. Figure 8.3 displays Tanzania's 19 districts, and Figure 8.4 gives Uganda's 55 districts. The administrative districts of Kenya are not used because several of them encompass different climate and vegetation zones.

The values of the Management Action node are actions that directly affect the ecosystem. The modeled actions are *poach for cash*, *poach for food*, *poach for protection*, *poach to protest*, *abandon settlement*, *indirectly damage wildlife habitat*, and *translocate animals*.

8.2.2   Habitat

Cheetah habitat is characterized by chance nodes for the region's climate (C), the proportion of a region's area that is protected (R_t), and how land is used that is not designated as a national park or a wildlife reserve (L). The proportion of protected land is allowed to change through time so that the creation and removal of reserves can be modeled.

8.2.3   Direct effect

A single direct effect on cheetah abundance is modeled: within-region poaching pressure (H_t). In future models, other direct effects such as disease could be added.

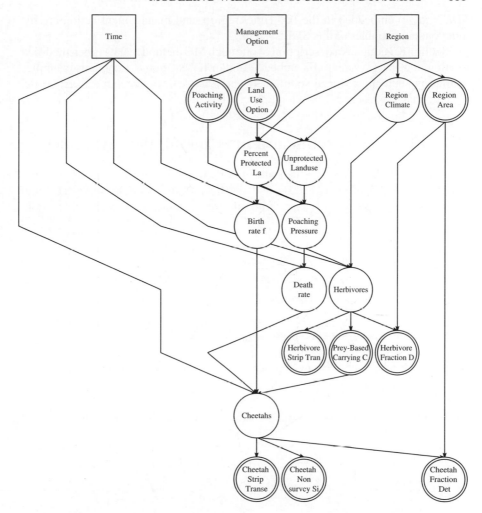

Figure 8.1   Ecosystem ID for the East African cheetah EMT.

8.2.4   Population dynamics

Cheetah population dynamics is modeled with a system of SDEs consisting of the within-region nodes of birth rate (f_t), death rate (r_t), herbivore abundance (B_t), cheetah carrying capacity (K_t), and cheetah abundance (N_t). This population dynamics model, then, is multivariate with dependent variables B_t, f_t, r_t, and N_t.

The herbivore abundance SDE is

$$\frac{dB_t}{dt} = \alpha_1 B_t(1 - B_t/\alpha_0) + \sigma dW_t \tag{8.1}$$

where α_0 is the habitat's herbivore carrying capacity, α_1 is the difference between herbivore birth and death rates, σ is the diffusion parameter, and W_t is a Wiener

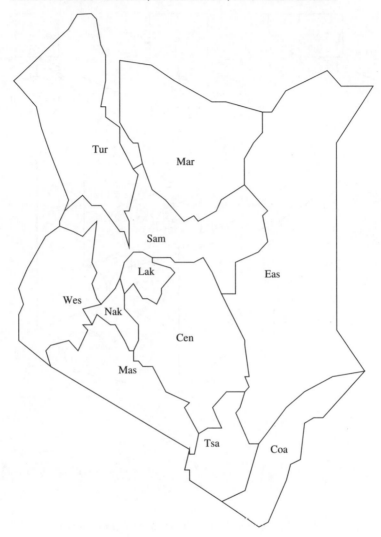

Figure 8.2   Kenya's regions of approximately homogeneous climate and vegetation regimes.

process.   The initial value is $B_0 = .6\alpha_0$. This model is a simplified version of the relationship given in Wells *et al.* (1998) derived by assuming that the probability is 1.0 that offspring will result from any union of a male herbivore and a female herbivore.

Converting a deterministic differential equation to an SDE by adding a term that is the derivative of a Wiener process is similar to how Manninen *et al.* (2006) modify a deterministic differential equation model of neuron operation into an SDE model.

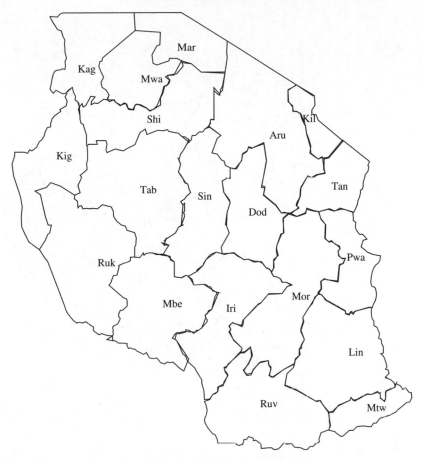

Figure 8.3   District boundaries for Tanzania.

As described in Haas (2001), the distribution of f_t at t is the solution to the SDE

$$df_t = -.5(\alpha_f + \beta_f^2(2f_t - 1))(1 - (2f_t - 1)^2)dt + .5\beta_f(1 - (2f_t - 1)^2)dW_t^{(f)}.$$
$$(8.2)$$

This SDE was chosen because its solution is bounded between 0 and 1, making f_t a dimensionless, fractional birth rate (fraction of abundance). A similar development for the death rate leads to the SDE

$$dr_t = -.5(\alpha_r + \beta_r^2(2r_t - 1))(1 - (2r_t - 1)^2)dt + .5\beta_r(1 - (2r_t - 1)^2)dW_t^{(r)}.$$
$$(8.3)$$

Note that the birth rate decreases as α_f becomes increasingly positive and the death rate decreases as α_r becomes increasingly positive.

Figure 8.4   District boundaries for Uganda.

The tendency of more females to have litters within protected areas (see Gros 1998) is represented by having the parameter α_f be conditional on the proportion of protected land in the region. Similarly, to represent the effect of poaching on r_t, α_r is conditional on poaching pressure (H_t). Poaching pressure, in turn, increases as (a) poaching activities increase, (b) the proportion of protected land decreases, and (c) land is used to keep livestock. The variability of the sample paths of f_t and r_t are controlled by the parameters β_f and β_r, respectively. Although the SDEs for f_t and r_t are not derived from biological theory, their use allows birth and death rates to be modeled as bounded, temporal stochastic processes with parameters that can represent different temporal trends (with $\alpha.$) and different amounts of variability (with $\beta.$).

Cheetah population dynamics is dependent on herbivore abundance through the cheetah carrying capacity, K_t. Following Gros (1998), this relationship is modeled as a linear function: $K_t = \beta_0 + \beta_1 B_t$.

Letting P, c, N_0 and β_N, be fixed parameters, the cheetah abundance SDE is

$$dN_t = \left[f_t(1 - P^{cN_t}) - r_t - (f_t - r_t)\frac{N_t}{K_t} \right] N_t dt + \beta_N dW_t^{(N)}. \qquad (8.4)$$

The parameter P is the probability that a meeting does not result in a litter, c is the proportion of animals that meet over a short time period, N_0 is the initial population size, and β_N is the noise coefficient. All unmodeled effects (such as migration/emigration and/or parameter values that are age dependent) that could influence the within-region cheetah abundance differential (dN_t) are represented by the derivative of a Wiener process ($dW_t^{(N)}$).

The solution to this or any SDE is a time-indexed probability distribution for the dependent variable (here, N_t). Using this distribution, the wildlife population has a *low extinction risk* as defined in Chapter 1 if $P(N_{t_F} > 5) > .01$ for $t_F = t_0 + 50$ *animal generations*.

To give some interpretation to this model, recall that the Malthusian growth model is $dN = (f - r)Ndt$, and the Pearl–Verhulst density-dependent growth model (logistic equation) is $dN = [(f - r) - (f - r)N/K]Ndt$. The first term in the cheetah abundance SDE is a simple birth rate effect adjusted for missed or barren unions, and the second term is a simple death rate effect. Hence, the first two terms in the cheetah abundance SDE make up the standard Malthusian growth model that has been modified to account for the chance of missed or barren unions. The third term in the cheetah abundance SDE is the Pearl–Verhulst addition to the Malthusian growth model to incorporate density-dependent population growth and has a negative effect when the birth rate exceeds the death rate – or a positive effect when the birth rate is smaller than the death rate. This effect, either way, is in proportion to the population size relative to the carrying capacity. For example, if the birth rate exceeds the death rate and the population size exceeds the carrying capacity, this density-dependent effect will be negative – one specific mechanism might be high infant mortality due to scarce food resources.

Malthusian population or reproduction kinetics are *first order* in that 'each individual in the population effectively results in the same number of offspring regardless of the population density,' (Wells *et al.* 1998) whereas, when the population density is high, P^{cN} will be small and the population is said to be governed by negative *second-order reproduction kinetics* (Wells *et al.* 1998). *Positive second-order reproduction kinetics* govern when P^{cN} is distinctly different from zero due to (a) low population density, (b) inefficient meeting mechanisms (a small value of c), or (c) reduced fecundity (a large value of P). An endangered species is vulnerable to entering a state of positive second-order reproduction kinetics (Wells *et al.* 1998).

Wells *et al.* (1998) develop this model through the pedagogical device of a finite difference equation but then go on to give a continuous-time version. This continuous-time version has been modified here to produce an SDE. The use of a

Table 8.1 Hypothesis values of parameters that define the population dynamics SDE system.

Node	Parents and values	Parameters and values			
B_t	Time, Rclimate, PchPrs	α_0	α_1	σ	
	time very_arid minor	2250.0	$-.0030$	.0001	
	time very_arid moderate	4500.0	.0140	.0005	
	time very_arid severe	8750.0	$-.0175$	.0002	
	time arid minor	7125.0	.0073	.0001	
	time arid moderate	7125.0	.0209	.0002	
	time arid severe	9500.0	.1179	.0002	
	time semi-arid minor	4500.0	.0181	.0001	
	time semi-arid moderate	9000.0	.0157	.0002	
	time semi-arid severe	9000.0	.1575	.0001	
	time non-arid minor	6562.5	.0393	.0001	
	time non-arid moderate	8750.0	.0787	.0014	
	time non-arid severe	8750.0	.4687	.0002	
K_t	B_t	β_0	β_1		
	abundance	.0	.1		
f_t	Time, PPtd	f_0	α_f	β_f	
	time low	.0448	.00236	.0016	
	time high	.0448	$-.00239$	.0030	
r_t	Time, PchPrs	r_0	α_r	β_r	
	time minor	.04480	$-.00003$	.0001	
	time moderate	.04480	$-.00009$	.0005	
	time severe	.04480	$-.00020$	.0070	
N_t	Time, f_t, r_t, K_t	N_0	P	c	β_N
	time rate rate abundance	200.0	.1	.3	.0023

continuous time model frees the modeler from having to justify a time increment and allows the model to be made stochastic through the well-developed theory of SDEs.

Of course, ecosystem health is incompletely characterized by herbivore and cheetah abundances. Improved models will need to include other animals and vegetation type such as fraction of area that is grassland.

8.2.4.1 Hypothesis values

Tables 8.1 and 8.2 give the hypothesis values of this model's parameters. Bashir *et al.* (2004) gives expert-judgment estimates that, in 2004, there were about 1200

Table 8.2 Hypothesis values of parameters that define areal fraction detection chance nodes of the ecosystem ID.

Node	Parents and values	Parameter values	
D_t^H	B_t, RArea	ξ^H	ρ^H
	abundance area	.0001	.9999
D_t^C	N_t, RArea	ξ^C	ρ^C
	abundance area	.0001	.9999

cheetahs in Kenya, 1000 in Tanzania, and 200 in Uganda. These values are used in Chapter 11 as hypothesis values of the expected value of the cheetah abundance node (N_t) in the year 2004.

8.2.5   Observable (output) random variables

The ecosystem ID's output nodes are based on B_t and N_t. Any stochastic model that is to have its parameters estimated from data needs to represent actual observations. To achieve this, the ecosystem ID's output nodes, B_t and N_t, are augmented with nodes that have distributions conditional on these two fundamental ones. These additional nodes are called *estimator nodes*.

Methods of gathering data on the abundance of an elusive and low-density population such as the cheetah's include transect surveys, nonsurvey sightings, camera traps, presence–absence surveys, areal detection fraction, hyper-spatial resolution satellite photographs, and spoor counts. An explanation is given in Chapter 10 for why the ecosystem ID needs to be able to represent several different ways of estimating animal abundance. Chapter 10 also contains more detail on these different abundance estimation methods.

The ecosystem ID of Figure 8.1 contains estimator nodes to represent transect surveys, areal detection fraction observations, and nonsurvey sightings only. This is because the data set of Chapter 10 for the East African cheetah EMT consists of observations on only these types of abundance estimators.

8.2.5.1   Transect survey node

This node represents transect survey counts, $Y_t = (a/A)N_t$ where $a = 2\mu L$ for strip and line transects, and equals $\pi \mu^2$ for fixed radius point counts and point transects (see Chapter 10, Section 10.3.1).

8.2.5.2   Nonsurvey sightings node

Let S_t be the number of nonsurvey sightings at time t. Let ρ be the number of animals per sighting. Then $S_t = N_t/\rho$. Because S_t is linear in N_t, it is a `Determ_Linear` type node (see Chapter 5, Section 5.3.3).

8.2.5.3   Camera trap node

Let Y_t be the number of photographs triggered by animals at time point t. Let C be a constant defined by parameters as described in Chapter 10, Section 10.3.3. Then $Y_t = CN_t$. Because Y_t is linear in N_t, it is a `Determ_Linear` type node.

8.2.5.4   Nodes for presence-absence data

Two animal abundance estimators are given in Chapter 10, Section 10.3.4, that make use of presence–absence data. The first one models presence–absence outcomes over a set of spatial cells with a multivariate Bernoulli distribution. The second estimator models the fraction of a region's area on which the animal is detected as a threshold function in animal abundance. These estimators can be represented in the ecosystem ID with the following two nodes.

Presence–absence nodes For $i = 1, \ldots, C$ where C is the number of cells, let S_{ti} be the value 1.0 if the animal is detected in cell i at time t, and 0 otherwise. Then, the joint probability of seeing $S_{t1} = s_{t1}, \ldots, S_{tC} = s_{tC}$ is given by Equation 10.4 in Chapter 10. All of the nodes $S_{t1}, \ldots, S_{tC}$ are of type `Multivariate_Bernoulli`.

Areal detection fraction nodes These output nodes are `Herbivore Detection Fraction` (D_t^H) and `Cheetah Detection Fraction` (D_t^C). These nodes measure the fraction of a region's area over which herbivores, and cheetahs respectively, have been detected. Let a_R be a region's surface area so that B_t/a_R is the density of herbivores in the region. An observation on D_t^H can be computed from maps of herbivore presence–absence by district. This is done by dividing the sum of all areas of districts in the region on which herbivores have been detected by a_R.

The ecosystem ID models D_t^H as a deterministic function of a_R and B_t as follows. Let ξ^H be the minimum herbivore density that results in a herbivore detection report. Let ρ^H be a herbivore density value above which herbivores are certain to be reported. Then

$$D_t^H = \frac{B_t/a_R - \xi^H}{\rho^H - \xi^H} I_{\{\xi^H < B_t/a_R < \rho^H\}}(B_t) + I_{\{\rho^H < B_t/a_R\}}(B_t). \qquad (8.5)$$

Note that it is possible for B_t to be positive but D_t^H to be zero, that is, ξ^H can be interpreted as the minimum density detection limit. D_t^C is defined in a similar manner.

8.2.5.5   Spoor count node

Let Y_t be the number of spoors (paw or hoof prints) at time t. Then $Y_t = CN_t$ where C is a constant defined by a set of parameters as described in Chapter 10, Section 10.3.6. Because Y_t is linear in N_t, it is a `Determ_Linear` type node.

8.2.6   Aggregating outputs across a country's districts

Because the ecosystem ID is conditional on region, herbivore and cheetah detection fractions are region specific. Since the group IDs are not regionally indexed, these region-specific ecosystem ID outputs need to be aggregated across regions. Here, this aggregation is accomplished by computing, at each time step, a weighted average of the expected values of ecosystem output nodes with region area as the weighting variable. These weighted averages for a particular country and time point are written to the bulletin board as an *ecosystem state message*. This message consists of the time, country, and expected values of D_t^H and D_t^C. These country-wide averaged outputs are then read by the group IDs as the single, aggregated value for that country at that time point.

8.3   Solving SDEs within an ID

Logic sampling (Henrion 1988) is used to approximate the marginal distribution of all nodes in the ID including SDE nodes. Logic sampling consists of simulating m realizations from the ID's joint distribution under a set of conditioning values on the ID's input nodes using the recursive factorization of the ID's joint distribution (see Kiiveri *et al.* 1984) to sample from each node's conditional distribution. For the ith such logic sample, the realized values of nodes that influence nodes that make up the system of SDEs are used as fixed values in one realization (or *pass*) of the SDE solver (described below) over the temporal interval on which the SDE is being solved. SDE node values at the end-time of this interval from this pass are used as parent values for the conditional distributions of all child nodes of these SDE nodes. Hence, there are m SDE solution passes through the SDE system and each such pass yields values for the logic sampling of non-SDE nodes that are children of SDE nodes.

8.3.1   Approximating the solution of a system of SDEs

Because the system of SDEs above does not have an analytical solution, it is solved from the last time point to the current time point using a numerical method. Specifically, the joint probability distribution of $(B_t, K_t, f_t, r_t, N_t)'$ conditional on the values of the input nodes is numerically approximated by generating a large number of sample paths of the SDE system with the *Explicit Order 1.0 Strong Scheme* described in Kloeden and Platen (1995, p. 376).

The kth equation in the SDE system is advanced one time step with

$$Y_{n+1}^k = Y_n^k + a^k \Delta + b^{k,k} \Delta W^k + \frac{1}{2\sqrt{\Delta}} \left\{ b^{k,k}(t_n, \tilde{\Upsilon}_n^k) - b^{k,k} \right\} \left\{ (\Delta W^k)^2 - \Delta \right\}$$

(8.6)

where $\tilde{\Upsilon}_n^k = Y_n + a\Delta + b^k\sqrt{\Delta}$, t_n is the nth time point, Δ is the time step, a^k is the drift function, $b^{k,k}$ is the diffusion function, and $\Delta W = W_{t_{n+1}} - W_{t_n}$

is the approximation to the Wiener process increment (see Kloeden and Platen 1995, p. 340).

8.4   Example of ecosystem ID output

Figure 8.5 displays the output from the East African cheetah EMT ecosystem ID. This solution is for the Kenya region of Tsavo under a constant management option of *poach for food* over the interval 1997 to 2060.

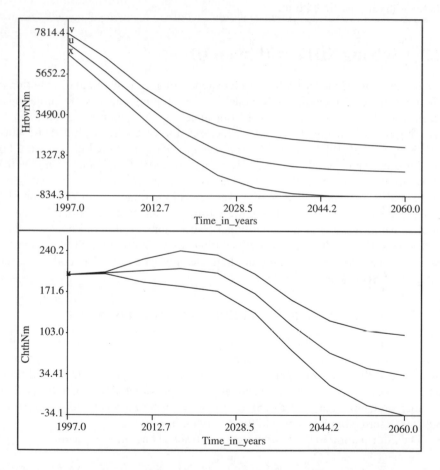

Figure 8.5   East African cheetah ecosystem ID output. The top plot shows the output on the B_t node, and the bottom plot the output on the N_t node. In each plot, the centerline is the mean, and the two bounding lines are the mean plus or minus one standard deviation. The region is Tsavo in Kenya and the management option of poach for food *is applied to the ecosystem ID at every time point.*

An essential characteristic of a wildlife population model is that its predictions exhibit increasing uncertainty as the prediction time extends into the future. As can be seen in Figure 8.5, an SDE naturally exhibits such an increase in uncertainty.

8.5   Conclusions

A way has been given for embedding, within an ID, a population dynamics model of wildlife abundance expressed as a system of SDEs. The ID architecture naturally allows nodes to be added to this model that represent several different forms of data that are useful for estimating animal abundance. This facility makes the ID-based ecosystem model well suited for use within a wildlife management agency that may have to scrounge and scavenge such data in order to support decision making for those species it is charged with managing.

The ecosystem ID also includes geographic variables that impact habit suitability. Such geographic variables, for example, landuse, can be developed and updated using **id**'s GIS capabilities.

Embedding an SDE system within an ID requires modification of the numerical methods that compute approximate SDE solutions. Such modifications are a part of the **id** software system.

To fix ideas and to show feasibility, output from the ecosystem ID for the East African cheetah EMT was presented.

8.6   Exercises

1. Modify parameter values of the ecosystem ID so that the model predicts the cheetah's extinction in 25 years.

2. Construct an ecosystem ID for the conservation of great white sharks (*Carcharodon carcharias*) along the east coast of the USA. Use the cheetah conservation ID of this chapter to get started.

3. Simplify the population dynamics SDE so that cheetah birth and death rates are nonrandom. Then, rerun the simulator over the time period 1997 to 2060. Comment on any differences in the output from this new population dynamics model relative to Figure 8.5, above.

9

Political action taxonomies, collection protocols, and an actions history example

9.1 Introduction

The group decision-making IDs have political action nodes. These nominally valued nodes have values that correspond to entries in a taxonomy of political actions. This chapter describes the taxonomy used to represent political actions that are related to ecosystem management.

The chapter's layout is as follows. Classification systems (taxonomies) for political actions are reviewed in Section 9.2 with an emphasis on the Behavioral Correlates of War (BCOW) taxonomy of Leng (1999). Section 9.3 contains the complete taxonomy used in the EMT that has enhanced abilities to represent actions that affect ecosystems. This taxonomy is constructed by modifying and extending the BCOW taxonomy and was first described in Haas (2008b). The coding protocol used to enter an action into an EMAT database is given in Section 9.4. Section 9.5 contains the sources of action observations used within the East African cheetah EMT (see Chapter 4, Figure 4.1). This section also contains a discussion of this data set. Conclusions are reached in the final section.

Improving Natural Resource Management: Ecological and Political Models   Timothy C. Haas
© 2011 John Wiley & Sons, Ltd

9.2   Political action taxonomies

A model that is to be used to predict what actions a group might take in the future needs to have a list of candidate actions rather than holding a list of unique, nonrepeatable actions that occurred in the past. One way to model such possible future actions is to find a list of action categories that categorize many possible future actions and then construct a model of how input actions drive these action categories. This approach needs a taxonomy of political actions (also referred to as a categorization scheme or classification system) that includes actions that involve ecosystem management. Such a taxonomy is developed in this chapter by extending an existing political actions taxonomy to represent one possible set of archetypal ecosystem management actions.

If groups are frequently inventing and executing new actions, a finite repertoire approach to modeling group decision making, as pursued here, may not be successful. One way to assess the usefulness of a finite repertoire action set is to see when/if actions repeat themselves. If such repetition is low, a different modeling approach may be more appropriate. An initial assessment of the prevalence of action repetition for the East African cheetah example is given in the last section of this chapter.

9.2.1   BCOW action taxonomy

Several action taxonomies or classification systems have been developed in the political science literature, see Schrodt (1995). From these taxonomies, the BCOW taxonomy of political actions (see Leng and Singer 1988) is used to develop a taxonomy of ecosystem management actions for two reasons. First, the BCOW taxonomy is designed to support a variety of theoretical viewpoints (Leng 1999) and hence can be used to code data that will be used to estimate a model of group decision making that, as is done here, synthesizes realist and cognitive processing paradigms of political decision making (see Chapter 6). Second, the BCOW taxonomy has coding slots for recording (a) a detailed description of an action, (b) inter- and intra-country groups, and (c) a short history of group interactions. The intention behind this last coding category is to allow causal relationships to be identified and tracked through time.

The BCOW taxonomy consists of a nearly exhaustive list of political actions grouped into militaristic, diplomatic, economic, unofficial (intra-country actor), and verbal categories. This taxonomy classifies actions taken by countries, and also by groups that are within countries. The first level in the BCOW taxonomy categorizes actions as either physical or verbal. A physical action is further classified as being militaristic, diplomatic, economic, or unofficial. Table 9.1, derived from Leng (1999), gives some of the BCOW codes and one-line descriptors of the physical actions that constitute the BCOW action typology. In this table, (D) indicates a discrete event with no temporal interval of occurrence, and (Reciprocal) indicates an event that involves an actor–target pair.

Table 9.1 Some BCOW actions and their codes.

Action	BCOW code
Military Actions	
Military victory (total)	11643
Military surrender	11621
Hostage taking	14153
Seizure	11423
Military victory (partial)	11633
Military intrusion	11443
Clash (reciprocal)	11513
Subversion or guerrilla activity	14113
Domestic military action (with verbal actions)	21719
Anti-foreign riot	14223
Commit atrocities	11673
Decrease in military trade relations	21121
Diplomatic actions	
Secede (D)	32153
Violate foreign territory	12233
Punish foreign representative(s)	12173
Violate international law	12223
Allow territorial political concession (D)	32611
Reach agreement (D) (Reciprocal)	12521
Negotiate conflict (Reciprocal)	12123
Change diplomatic relationship	12161
Anti-foreign demonstration	14213
Domestic political action (with verbal acts)	12719
Attend international event	12631
Economic actions	
Confiscate or nationalize	23163
Pay reparations (D)	23241
Pay for goods or services (D)	23231
Freeze foreign assets	23153
Sell or trade (D)	23121
Increase in economic assistance	23141
Economic grant (D)	23111
Expand in trade relations	23151
Reach economic agreement (D) (Reciprocal)	13551
Provide humanitarian aid or relief	23301
Domestic economic action (with verbal acts)	23719
Loan (D)	23131
Economic negotiation (Reciprocal)	13121
Shrink in trade relations	23151

(continued)

Table 9.1 (*Continued*)

Action	BCOW code
Unofficial actions	
Subversion or guerrilla activity	14113
Assassinate	14143
Consult	14151
Kidnap or hold hostage	14153
Anti-foreign demonstration	14213
Anti-foreign riot	14223
Pro-foreign demonstration	14251
Seize property	14263
Unofficial domestic action (with verbal act)	14719

Verbal actions are associated with some of these physical actions. Such actions do not form a separate group of types but rather point to a past, current, or future physical action. Hence, verbal actions do not have codes or one-line descriptors.

9.2.1.1   BCOW data structure

An observed action is stored in a BCOW database as follows. First, the action is identified by its date, actor, target, and location. A physical action is further identified by its action code.

A verbal action, on the other hand, is further identified by the BCOW code of the physical action that it is referring to, and by the type of that reference. These reference types are: (a) Comment-on-an-Action, (b) Request-an-Action, and (c) Intend-an-Action. If the referent action is a member of the database, this action's database identification number is added to the list of this verbal action's identifiers.

9.3   Adapting the BCOW taxonomy to ecosystem management actions

In this section, the BCOW taxonomy is modified and extended to produce a taxonomy that captures inter- and intra-country political actions that affect ecosystems. This new taxonomy is called the *Ecosystem Management Actions Taxonomy* (EMAT). EMAT does not contain the `Unofficial Actions` category of the BCOW taxonomy because, in EMAT, groups internal to a country are modeled as having nearly the same range of actions as a country-level group. Therefore, all BCOW unofficial actions are absorbed within EMAT into one of the other action categories.

EMAT contains action categories that are not present in the BCOW taxonomy. The sources of such actions are (a) actions peculiar to ecosystem management,

for example, `Translocate_animals`, (b) the need for subcategories of
BCOW action categories, and (c) actions taken by the ecosystem, for example,
`Lions_kill_rural_residents_and_livestock`. An example of source
(b) is the BCOW action category `Seizure`. In the political–ecological system
simulator, groups are sensitive to the seizure of different things: for example,

$$\texttt{Seize\_elephant\_tusks}$$

is different than

$$\texttt{Seize\_private\_land\_to\_give\_to\_the\_poor}.$$

The third source is added to EMAT because the BCOW taxonomy, having been
developed (mainly) to study actions between international political entities (countries), contains only actions taken by anthropogenic actors. Actions in this third
category are given at the end of the EMAT listing and have codes beginning with
the letter Z.

EMAT inherits the BCOW taxonomy's definition of a military action: an
action towards another group that involves violence. Hence, in EMAT, although
the shooting or otherwise killing of animals is violent, these are not actions
towards another group. Anti-poaching actions, however, being violent actions
towards another group, are categorized as military actions.

EMAT preserves the original BCOW codes by using only a single leading letter.
For such an action, the EMAT code has the following format:

`Action Type (M, D, or E) BCOW Code Xii (where ii is a 2-digit number).`

This format forces the BCOW code matching routine in **id** to not match on an
original BCOW action category such as `Military_Victory_[Total]`
because these `[...]` tags are not included in any of the group ID action lists.
See, for example, action category M11513: `Attack:_general_type`. Categories that have been added to the BCOW taxonomy, however (all codes beginning
with the letter C or double letters), repeat the given code for special cases. See, for
example, action category MM0008: `Evict_residents_from_reserve`.

9.3.1   EMAT listing

The following is the complete listing of EMAT categories with associated codes:

```
Military_Victory_[Total]  M11643
Military_Surrender  M11621
Hostage_Taking  M14153
Take_Prisoners_of_War  M11663
   Abduct_some_residents  M11663
Blockade  M11433
Seizure:_general_type_[inter-country] M11423
```

```
Military_Withdrawal  M11521
Military_Victory_[Partial]  M11633
Attack:_general_type  M11523
   Attack_and_rob_some_tourists  M11523X1
   Murder_tourists  M11523X1
   Murder_pastoralists_and_steal_cattle M11523X1
   Murder_some_villagers  M11523X1
   Murder_some_residents  M11523X1
   Violently_attack_pastoralists  M11523X1
   Murder_some_game_wardens  M11523X1
      Kill_wildlife_personnel  M11523X1
Assassination_[D]  M14143
Military_Occupation  M11653
Military_Intrusion  M11443
Enter_Demilitarized_Zone  M11453
Fortify_Occupied_Territory  M31133
Clash_[Reciprocal]  M11513
   Violently_clash_with_pastoralists_over_water_access M11513X1
   Murder_some_residents_over_water_access M11513X1
   Engage_in_ethnic_riots  M11513X1
Subversion_or_Guerrilla_Activity  M14113
Continuous_Military_Combat  M11533
Mobilization  M11353
Show-of-Force_[or_Maneuvers]  M11313
Domestic_Military_Action_[With_Verbal_Actions]:_general_type M21719
   Suppress_riot  M21719X1
   Riot_suppression  M21719X1
   Kill_some_insurgents  M21719X1
   Harass_and_intimidate_pastoralists M21719X1
   Extort_pastoralists  M21719X1
Antiforeign_Riot  M14223
Commit_Atrocities  M11673
Grant_or_Sell_Nuclear_Weapons  M21311
Alert  M11333
Remove_Nuclear_Weapons  M21333
Remove_Military_Things  M21233
Return_Military_Things  M21211
Increase_in_Combat_Force_Level  M21143
Increase_Force_Level  M21133
Military_Grant  M21111
Establish_Military_Base  M31132
Lease_Territory  M31111
Internal_Peacekeeping_Force  M11212
Military_Advice_or_Assistance  M11121
Military_Coordination  M11131
Increase_in_Military_Trade_Relations M21121
Permit_Foreign_Military_Presence  M31121
Increase_in_Military_Assistance  M21141
Decrease_Force_Level  M21133
Decrease_in_Combat_Force_Level  M21143
Decrease_in_Military_Assistance  M21141
Decrease_in_Military_Trade_Relations M21121
Antiguerrilla_Action  M11363
```

```
Evict_residents_from_reserve   MM0008
   Mass_eviction_from_reserve   MM0008
   Mass_eviction_from_wildlife_migration_routes MM0008
   Eviction_from_forest   MM0008
   Evict_pastoralists_from_wetlands MM0008
   Evict_some_RRs_from_public_lands MM0008
Drive_livestock_out_of_NP   MM0009
Increase_antipoaching   MM0010
Increase_anti-wildlife_trade_enforcement MM0011
Decrease_antipoaching   MM0012
Recruit_poachers_to_be_antipoaching_police MM0013
Detain_some_RRs_for_forest_destruction MM0014
   Detain_some_RRs_for_encroachment_and_tree_destruction MM0014
Arrest_some_RRs_for_poaching   MM0015
   Arrest_ivory_poachers   MM0015
   Shoot_some_RRs_for_poaching   MM0015
   Shoot_some_EPA_officers_for_poaching MM0015
Arrest_some_EPA_officers_for_poaching MM0016
   Arrest_some_EPA_officers_for_extortion MM0016
Surrender_poaching_weapons   MM0017
Antigovernment_Riot   MM0018
   Violently_clash_with_political_rivals M00018
   Riot_over_grazing_land_rights   MM0018
   Riot_over_land_allocation   MM0018
   Riot_to_protest_eviction   MM0018
   Attack_wardens_to_resist_eviction MM0018
   Protest_by_blocking_highway   MM0018
   Drive_cattle_to_protest_land_seizure MM0018
   Refuse_order_to_vacate_settlement MM0018
   Riot_to_acquire_hippo_bushmeat MM0018
Comment ccccccccccccccccccccccccccccccccccccccccccccccccccccccccccccccc
Secede_[D]   D32153
Violate_Foreign_Territory   D12233
Punish_or_Restrict_Foreign_Nationals D12213
Punish_Foreign_Representative[s]   D12173
Violate_International_Law   D12223
Grant_Independence_[D]   D32151
Grant_or_Cede_Territory_[D]   D32111
Allow_Territorial_Political_Concession_[D] D32611
Assume_Foreign_Throne_[D]   D12641
Annex_[D]   D32143
Peace_Settlement_[D]_[Reciprocal] D12361
Integrate_[D]_[Reciprocal]   D32161
Partition_[D]   D32173
Colonize_or_Claim_Territory_[D]   D32111
Establish_Protectorate_or_Trusteeship_[D] D32142
Declare_Neutrality_[D]   D12362
Declare_Independence_[D]   D32163
Form_Alliance_[D]_[Reciprocal]   D12511
Break_Diplomatic_Agreement_[D]   D12533
Break_Off_Talks_or_Negotiations_[D] D12123
Guarantee_Agreement   D12152
Reach_Agreement_[D]_[Reciprocal]   D12521
```

```
    Sign_inter-country_conservation_accord D12521X1
    Ratify_Convention_on_Biological_Diversity D12521X1
    Sign_inter-country_tourism_protocol D12521X1
    Agree_to_form_a_cross_border_national_park D12521X1
    Negotiate_trans-border_antirustling_enforcement_plan D12521X1
    Negotiate_a_several_year_ivory_ban_extension D12521X1
    Sign_agreement_to_create_tourism_school D12521X1
Intervene_Politically  D12243
Increase_in_Border_Restrictions  D12183
    Decrease_in_Border_Restrictions D12183
Establish_Demilitarized_Zone_[D]  D32141
Adjudicate  D12352
Declare_War_[D]  D12363
Provide_Diplomatic_Assistance  D12131
Mediate_or_Arbitrate  D12142
Negotiate_Conflict_[Reciprocal]  D12123
    Dispute_tourism_fee_collection D12123X1
    Dispute_NP_expenditures  D12123X1
    Sue_over_breach_of_job_contract D12123X1
Conduct_Plebiscite  D12232
Consult_[Reciprocal]  D12111
    Negotiate_for_a_single_tourist_visa D12111X1
Change_Diplomatic_Relationship  D12161
Proforeign_Demonstration  D14251
Antiforeign_Demonstration  D14213
Investigate  D12342
    Investigate_corruption_allegations D12342X1
    Investigate_endangered_plant_smuggling D12342X2
Domestic_Political_Action_[With_Verbal_Acts] D12719
    Lobby_to_legalize_sport_hunting D12719X1
    Negotiate_increased_tourism_capacity D12719X1
    Petition_to_stop_wildlife-caused_crop_destruction D12719X1
    Stop_antelope_translocation  D12719X1
    Lobby_to_stop_black_rhino_culling D12719X1
Change_in_Diplomatic_Representation D12151
Attend_International_Event  D12631
Grant_Asylum  D12373
Express_dissatisfaction_with_policy DD0001
    Demand_higher_compensation_for_wildlife-caused_damage DD0001
    Protest_tourism_compensation  DD0001
    Request_higher_percentage_of_NP_fees DD0001
    Protest_wildlife_export_plan  DD0001
    Protest_NP_licenses  DD0001
Verbal:_Intend_Action  DD0002
    Threaten:_block_roads_over_debt_dispute DD0002X1
Verbal:_Request_Action:_general_type DD0003
    Request_increased_antipoaching DD0003X1
    Request_ivory_trade_ban_continuation DD0003X1
    Request:_make_species_a_protected_species DD0003X1
    Request:_do_not_support_wildlife_trade DD0003X1
    Request_investigation_of_excessive_hunting DD0003X1
    Request:_protect_species_habitat DD0003X1
    Request:_survey_and_mark_NP_boundaries DD0003X2
```

```
   Request:_stop_human-wildlife_conflict DD0003X2
   Request:_remove_marauding_lion DD0003X2
   Declare:_will_kill_crop-damaging_elephants DD0003X2
   Request_to_graze_inside_NP  DD0003X2
   Request:_commercialize_elephant_hunting DD0003X2
   Offer:_will_stop_poaching_for_livelihood DD0003X2
Impose_resource-use_ban  DD0004
   Impose_some_months_logging_ban DD0004
   Bar_rural_residents_from_farming_land DD0004
   Stop_wildlife_export_plan  DD0004
Pass_tourism_industry_stimulus_legislation DD0005
   Give_tax_incentives_to_encourage_tourism_investment DD0005
Pass_environmental_protection_legislation DD0006
   Pass_wetlands_protection_legislation DD0006
   Set_water_pollution_rules  DD0006
Award_EPA_personnel  DD0007
   Award_some_EPA_personnel_for_exemplary_service DD0007
   Give_award_to_park_rangers  DD0007
   Award_some_rangers_for_exemplary_service DD0007
Verbal:_Comment_on_Action  DD0008
   Negative_ecoreport  DD0008X1
      Report:_wetlands_heavily_degraded DD0008X1
      Report:_grasslands_heavily_degraded DD0008X1
      Report:_human_activities_destroying_forest_belt DD0008X1
      Report:_region_being_devastated DD0008X1
      Report:_wildlife_flee_human_encroachment DD0008X1
      Report:_possible_lion_poisoning DD0008X1
      Report:_mass_dieoff_of_species DD0008X1
      Report:_species_population_is_decreasing_in_NP DD0008X1
      Report:_disease_endangers_lions DD0008X1
      Report:_forest_fragmentation_fosters_monkey_disease DD0008X1
      Report:_lake_is_drying_up  DD0008X1
      Report:_China_driving_ivory_boom DD0008X1
      Report:_much_of_area_meat_is_bushmeat DD0008X1
      Report:_civil_war_hurts_tourism DD0008X1
   Positive_ecoreport  DD0008X2
      Report:_census_shows_species_population_increase DD0008X2
      Report:_elephant_count_shows_increase DD0008X2
      Report:_new_species_found  DD0008X2
      Report:_new_bird_species_found DD0008X2
      Report:_new_monkey_species_found DD0008X2
      Report:_wildlife_conservation_helps_sustainable_development DD0008X2
      Declare_national_tree_planting_day DD0008X2
      Report:_census_of_a_species DD0008X2
      Report:_only_shoot_lions_older_than_five_years DD0008X2
   Verbally_protest_NP_boundaries DD0008X3
   Send:_letter_to_protest_NP_boundaries DD0008X3
   Report:_national_tourism_business_is_breaking_even DD0008X4
Punish_or_restrict_corrupt_ministers DD0009
   Ban_travel_of_some_corrupt_ministers DD0009
   Prosecute_corrupt_officials  DD0009
   Prosecute_embezzlers  DD0009
   Fire_corrupt_environmental_manager DD0009
```

```
   Fire_or_suspend_corrupt_EPA_officers DD0009
   Suspend_corrupt_forestry_officers DD0009
   Suspend_some_corrupt_officials DD0009
Resign_from_Chief_of_EPA_position DD0010
Receive_resignation_as_protest_against_condition DD0011
Appoint_new_EPA_director   DD0012
Tighten_wildlife_agreement_or_laws DD0013
   Order_local_govs_to_legislate_against_poaching DD0013
   Veto_pro-hunting_bill   DD0013
   Revoke_hunting_rights   DD0013
   Legalize_existing_private_reserves DD0013
   Refuse_reserve_privatization_request DD0013
   Freeze_new_facilities_construction DD0013
   Reject_NP_development_request   DD0013
   Stop_rainforest_sale_to_developer DD0013
Relax_wildlife_agreement_or_laws   DD0014
   Modify_Wildlife_Act_to_compensate_victims DD0014
   Remove_ivory_ban   DD0014
   Legalize_hunting_on_private_land DD0014
   Lighten_penalty_for_endangered_species_trade DD0014
   Increase_number_of_gorilla_tracking_permits DD0014
   Lift_hunting_ban   DD0014
   Allow_honey_industry_in_NP   DD0014
   Suspend_forest_evictions   DD0014
Host_or_attend_conservation_conference DD0015
   Sponsor_conference_on_mountain_area_conservation DD0015
   Attend_sustainable_conservation_conference DD0015
   Participate_in_conference_on_river_management DD0015
Host_or_attend_tourism_conference DD0016
Attend_tourism_conference_in_a_place DD0016
Agree_to_create_wildlife_reserves DD0017
Elect_environmentalist_legislator DD0018
Comment ccccccccccccccccccccccccccccccccccccccccccccccccccccccccccccccccc
Pay_Reparations_[D]   E23241
Pay_Ransom_[D]   E23251
Pay_Tribute_[D]   E23261
Pay_for_Goods_or_Services_[D]   E23231
   Acquire_equipment_for_wildlife_relocation E23231X1
Freeze_Foreign_Assets   E23153
Allow_Economic_Concession_[D]   E33111
Permit_Foreign_Economic_Passage_[D] E33131
Economic_Intrusion   E33133
   Steal_food_and_crops   E33133X1
   Steal_livestock   E33133X1
Sell_or_Trade_[D]:_general_type   E23121
   Trade_equipment_for_natural_resources E23121X1
   Offer_medical_equipment_for_timber E23121X1
   Export_some_monkeys   E23121X1
   Import_some_lions_to_zoo   E23121X1
   Import_some_rhinos   E23121X1
Increase_in_Economic_Assistance   E23141
Expand_in_Trade_Relations:_general_type E23151
   Begin_inter-country_economic_development_project E23151X1
```

```
   Begin_inter-country_energy_development_project E23151X1
   Begin_joint_biotechnology_venture E23151X1
   Engage_in_joint_research_projects E23151X1
Shrink_in_Trade_Relations   E23151
   Cancel_antipoaching_patrol_vehicle_purchase E23151X1
Reach_Economic_Agreement_[D]_[Reciprocal] E13551
   Sign_inter-country_customs_pact E13551
Provide_Humanitarian_Aid_or_Relief E23301
Economic_Integration_[D]_[Reciprocal] E23171
Economic_Coordination_[Reciprocal] E13211
Economic_Consultation_[Reciprocal] E13111
Loan_[D]:_general_type  E23131
   Loan_approval:_some_dollars_for_roads E23131X1
   Loan_some_dollars_for_roads  E23131X1
Economic_Negotiation_[Reciprocal] E13121
Decrease_in_Economic_Assistance:_general_type E23141
   Withhold_aid_due_to_corruption_concerns E23141X1
   Suspend_anti-corruption_aid  E23141X1
Advise_[Economic]  E13131
Domestic_Economic_Action_[With_Verbal_Acts]:_general_type E23719
   Fund_rural_development_project E23719X1
      Begin_a_trade_and_economic_development_trust E23719X1
      Fund_arid_land_development  E23719X1
      Build_fences_to_rehabilitate_drylands E23719X1
      Fund_elephant_control_trenches E23719X1
      Plan_water_storage_upgrades E23719X1
      Fund_watershed_management_project E23719X1
      Complete_road_building_project E23719X1
   Increase_wildlife_damage_control_measures E23719X2
      Complete_electric_wildlife-control_fence E23719X2
      Build_wildlife-control_fence E23719X2
      Hold_benefit_luncheon_for_lion-caused_livestock_loss_compensation
      E23719X2
   Fund_wildlife_management_education_grant E23719X3
      Fund_conservation_education E23719X3
      Fund_ecology_research_center E23719X3
   Fund_conservation_project  E23719X3
      Build_wildlife_migration_corridor E23719X3
      Pay_residents_to_protect_wildlife E23719X3
      Pay_residents_to_protect_turtle_nests E23719X3
      Train_some_new_antipoaching_rangers E23719X3
   Begin_project:_tree_planting  E23719X3
      Begin_project:_acacia_tree_planting E23719X3
      Begin_campaign:_agroforestry_tree_planting E23719X3
      Begin_campaign:_tree_planting_education E23719X3
      Plant_some_trees  E23719X3
   Create_a_private_wildlife_reserve E23719X4
      Convert_ranch_to_wildlife_reserve E23719X4
      Privatize_game_reserve  E23719X4
      Run:_sport_hunting_tourism  E23719X4
   Plan_tourism_infrastructure_improvement E23719X6
      Plan_to_build_NP_hotels  E23719X6
      Plan_to_build_tourist_lodge E23719X6
```

```
    Plan_to_build_NP_bridges_and_roads E23719X6
    Plan_to_construct_reserve_barriers E23719X6
  Invest_in_tourism_infrastructure E23719X7
    Train_tourist_guides  E23719X7
    Begin_campaign_to_encourage_tourism E23719X7
    Start_project_to_increase_avi-tourism E23719X7
    Launch_NP_promotion_campaign E23719X7
    Create_tourism_protection_police_force E23719X7
    Form_anti-wildlife_crime_task_force E23719X7
    Build_NP_facilities  E23719X7
    Upgrade_NP_facilities  E23719X7
    Upgrade_tourism_facilities  E23719X7
    Build_a_NP_airstrip  E23719X7
    upgrade_NP_airstrips  E23719X7
    Build_a_NP_airstrip_and_bridge E23719X7
    Build_tourist_lodges  E23719X7
    Rebuild_fence_around_reserve E23719X7
    Start_rhino_viewing_facility E23719X7
  Increase_NP_user_fees  E23719X8
    Raise_NP_leases  E23719X8
  Distribute_some_tourism_revenue_to_RRs E23719X9
    Distribute_some_NP_revenue_to_RRs E23719X9
  Experience_financial_shortfall E23719X10
    Experience_some_tourism_loss_due_to_wildlife_exodus E23719X10
    Experience_tourism_decrease E23719X10
  Experience_some_percent_NP_revenue_increase E23719X11
    Experience_some_tourism_revenue_increase E23719X11
    Experience_NP_visitor_increase E23719X11
  Increase_EPA_funding_by_some_percent E23719X12
  Reach_public-private_economic_agreement E23719X13
    Pay_commission_to_park-visit_organizers E23719X13
    Negotiate_NP_concessions  E23719X13
  Erode_soil_through_excessive_cultivation E23719X14
  Enter_a_refugee_camp_in_large_numbers E23719X15
  Drown_from_flood  E23719X16
  Experience_property_damage_from_heavy_rains E23719X17
  Starve_due_to_drought  E23719X17
    Experience_famine_due_to_drought E23719X17
    Lose_some_cattle_due_to_drought E23719X17
    Experience_severe_food_shortages E23719X17
  Open_gold_mine  E23719X5
Economic_Grant_[D]:_general_type  E23111
  Donate_to_project:_construct_reserve_barriers E23111X1
    Donate_some_dollars_for_rhino_fence E23111X1
  Donate_some_dollars_for_rural_development_project E23111X1
    Invest_some_dollars_in_rural_development_project E23111X1
    Donation_to:_rural_development_project E23111X1
    Donate_some_dollars_to_eradicate_rinderpest E23111X1
    Donate_some_dollars_to_conservation_through_rural_development E23111X1
    Donate_some_dollars_to_arid_land_development E23111X1
    Donate_some_dollars_for_water_storage_upgrades E23111X1
    Donate_some_dollars_for_irrigation_projects E23111X1
    Donate_some_dollars_for_roads E23111X1
```

```
    Donate_some_dollars_to_conservation_project E23111X1
        Donate_some_dollars_for_ecosystem_management E23111X1
        Donate_some_dollars_to_build_NP_improvements E23111X1
        Donate_some_dollars_for_wildlife_capture_equipment E23111X1
    Donate_some_dollars_for_antipoaching_patrols E23111X1
        Donate_money_for_black_rhino_antipoaching E23111X1
    Donate_antipoaching_equipment  E23111X1
        Donate_data_acquisition_equipment E23111X1
        Donate_some_dollars_for_communication_system E23111X1
        Donate_radios_for_antipoaching_patrols E23111X1
        Donate_some_vehicles_for_antipoaching_patrols E23111X1
        Donate_some_ivory_detectors_for_antipoaching_patrols E23111X1
        Donate_equipment_for_antipoaching_enforcement E23111X1
    Donate_bicycles_to_replace_those_destroyed_by_wildlife E23111X1
    Donate_some_farm_tools_to_reformed_poachers E23111X1
    Donate_to_establish_wildlife_trust_fund E23111X1
    Donate_for_wildlife_management_education_grant E23111X1
        Sponsor_veterinarian_training_program E23111X1
    Donate_some_dollars_for_tourism_marketing E23111X1
Open_reserve_to_settlement  C0004
    Open_reserve  C0004
    Reduce_game_reserve_size  C0004
    Sell_grazing_land_to_individuals C0004
Abandon_settlement  C0005
Indirectly_damage_wildlife_habitat C0006
    Destroy_lake_from_farming_and_industry C0006
        Reduce_lake_level_through_excessive_usage C0006
        Dry_up_lake_through_deforestation C0006
    Over-use_roads_and_cause_bird_population_decline C0006
    Indirect_action:_electrocute_birds_on_transmission_lines C0006
        Indirect_action:_kill_some_turtles_with_nets C0006
    Block_migration_routes_with_settlements C0006
        Build_houses_on_migration_path C0006
    Scare_away_elephants_with_translocation_helicopter C0006
    Dry_up_reservoir_with_sand_harvesting C0006
    Perform_illegal_logging  C0006
    Clear_new_land  C0006
        Clear_new_land_and_fell_trees C0006
    Burn_wetlands_and_displace_wildlife C0006
        Destroy_wetlands_through_development C0006
    Degrade_forest_through_over-grazing_and_tree_removal C0006
        Destroy_forest_through_burning_and_tree_removal C0006
    Degrade_watershed_through_mining_activities C0006
Confiscate_or_Nationalize:_general_type E23163
    Create_a_new_NP  E23163X1
        Convert_some_forests_to_NPs E23163X1
        Create_new_wildlife_reserve E23163X1
        Expand_rhino_sanctuary_boundaries E23163X1
        Open_rhino_sanctuary  E23163X1
        Merge_reserve_with_NP  E23163X1
    Seize_elephant_ivory  E23163X1
        Seize_elephant_tusks  E23163X1
        Confiscate_ivory_shipment  E23163X1
```

```
    Seize_wildlife_from_poachers  E23163X1
       Seize_crate_of_smuggled_wildlife E23163X1
       Rescue_chimpanzees_from_poachers E23163X1
       Seize_bushmeat  E23163X1
    Seize_stolen_forest_products  E23163X1
    Seize_idle_private_land_for_the_poor E23163X1
       Repossess_illegally_taken_land E23163X1
    Impound_cattle  E23163X1
Use_technology_to_find_wildlife_habitat C0007
    Use_mobile_phones_to_find_wildlife_habitat C0007
    Install_tracking_devices_on_wildlife C0007
    Upgrade_habitat-management_computer_systems C0007
Translocate_animals  C0008
    Move_animals  C0008
    Translocate_crocodiles  C0008
    Relocate_marauding_crocodile  C0008
    Translocate_elephants  C0008
    Move_elephants  C0008
    Move_elephants_from_region_to_reserve C0008
    Translocate_rhinos  C0008
    Move_rhinos  C0008
    Introduce_rhinos_into_region  C0008
    Introduce_black_rhinos_into_region C0008
    Introduce_white_rhinos_into_region C0008
    Introduce_antelopes_into_region C0008
    Rescue_lion_cubs  C0008
    Rescue_python  C0008
Kill_marauding_wildlife  C0009
    Shoot_marauding_elephants  C0009
    kill_man-eating_lion  C0009
Inoculate_wildlife_against_anthrax C0010
Conduct_de-snaring_and_resuscitation_patrols C0011
Guard_crops_against_marauding_wildlife C0012
Poach_for_cash  C0013
    Shoot_lions  C0013
    Kill_lions  C0013
    Shoot_elephants  C0013
    Kill_clephants  C0013
    Shoot_rhinos  C0013
    Kill_rhinos  C0013
    Shoot_black_rhinos  C0013
    Kill_black_rhinos  C0013
    poach_some_white_rhinos  C0013
Poach_for_food  C0014
    Shoot_animals  C0014
    Kill_animals  C0014
    Increase_use_of_snare_traps  C0014
Poach_for_protection  C0015
    Poison_wildlife_to_protect_livestock C0015
    Poison_some_leopards_to_protect_livestock C0015
    Poison_some_lions_to_protect_livestock C0015
    Poison_some_elephants_to_protect_crops C0015
Poach_to_protest  C0016
```

```
Poach_wildlife_for_rituals   C0017
   Poach_lions_for_rituals   C0017
Smuggle_wildlife_out_of_country   C0018
   Smuggle_birds_across_border   C0018
Legally_harvest_wildlife   C0019
   Collect_eggs_for_crocodile_farm C0019
Comment ccccccccccccccccccccccccccccccccccccccccccccccccccccccccccccccccc
Ecosystem_state   Z001
Disease_outbreak_kills_wildlife   Z002
   Disease_outbreak_kills_some_hippos Z002
   Mass_dieoff_of_flamingos   Z002
   Drought_induced_anthrax_kills_zebras Z002
   Post-drought_bloating_kills_wildlife Z002
   Disease_kills_some_rhinos   Z002
Flamingos_die_due_to_shrinking_lake Z003
Flamingos_population_increases   Z004
Elephants_trample_to_death_some_rural_residents Z005
   Elephants_trample_to_death_some_tourists Z005
Elephants_trample_crops   Z006
   Elephants_trample_crops_and_force_school_closure Z006
Elephants_trample_rural_residents_and_crops Z007
Elephants_trample_crops_and_livestock Z007
Wildlife_attack_rural_residents   Z008
   Wildlife_attack_rural_residents_for_food Z008
   Wildlife_attack_rural_residents_and_livestock Z008
   Lions_kill_rural_residents_and_livestock Z008
   Lions_maul_some_humans   Z008
   Cheetah_mauls_some_humans   Z008
   Leopard_mauls_some_humans   Z008
   Elephant_mauls_some_tourists   Z008
   Chimpanzees_maul_some_humans   Z008
   Baboon_kills_dogs_and_uproots_crops Z008
   Hyenas_kill_some_humans   Z008
   Hyenas_maul_some_goats   Z008
Wildlife_counts_increase   Z009
   Rhino_gives_birth   Z009
Reintroduced_antelope_die   Z0010
```

For the East African cheetah management application, there are some action categories in EMAT that happen infrequently. No attempt has been made, however, to identify and delete such actions since doing so might result in the deletion of an action category that is recurrent in some unforeseen future application of this tool.

9.4   EMAT coding protocol

9.4.1   Coding manual for group actions data

Record actions into an ASCII file using the following protocols.

Table 9.2 Fields for coding an observed action for entry into an EMAT database.

Field	Field content
1	Document Archive Number (DAN)
2	Story date (month day year)
3	Source of story
4	Number of actors
5	List of actors
6	Number of subjects
7	List of subjects
8	Action phrase
9	Number of countries subjected to action
10	List of countries
11	Number of regions
12	List of regions
13	Date of action

9.4.1.1   Protocol for acquiring action observations:

1. Search for news stories by searching for strings that relate to the eco-system being managed in one of the News Search Engines found on the data_sets page at www4.uwm.edu/people/haas/cheetah_emt. For example, with the East African cheetah EMT, one searches for strings that have one or more of the following phrases: '*country* cheetah,' '*country* wildlife,' '*country* national parks,' '*country* poaching,' '*country* poachers,' or '*country* land management,' where *country* is Kenya, Tanzania, or Uganda. Avoid opinion or 'study' stories. Next, search individual news story source websites using these same search phrases. Print or otherwise store each story so found. Underline relevant content in each of these stories.

2. Read each story found in the previous step and create an entry in the group actions history database by filling in the fields given in Table 9.2. Use a space to separate fields, components of a list, and components of a date.

Protocol for creating an action phrase:

1. Use only alpha numeric characters.

2. Express numbers numerically rather than with words, for example, enter 4_antelopes_die instead of four_antelopes_die. If unit or scale designation is needed, enter a number followed by the unit or scale, for example, km2, K, or M. Some examples of these conventions are:

`4_rhinos_translocated`, `500km2_forest`, `5K_dollars`, and `13.4M_dollars`.

3. Express an action in the present tense, for example, `award_personnel` instead of `awarded_personnel`.

4. Construct declarative (active voice) phrases. In particular, place the action's verb as close as possible to the beginning of the phrase, e.g. **move_400_elephants** instead of **400_elephants_are_moved**.

5. Use the word `reserve` instead of 'preserve' when referring to wildlife reserves (preserves) and/or wildlife sanctuaries. See, for example, the website `www.game-reserve.com`.

Protocol for adding a category to EMAT:

Add EMAT categories as necessary to the EMAT definition file (*emat.dfn*) so that each observed action belongs to exactly one EMAT category. Add such a new category as follows:

1. Print out the file *noemat.err* that was produced by running the **prepare_raw_group_data** relation in **id**.

2. Open two windows on the computer screen. In one of these windows, edit the EMAT definition file and edit the observed actions data file in the other window.

3. Using these two windows, search for the verb of an unrecognized observed action in the EMAT definition file to navigate to the relevant area of the EMAT definition file.

Using this technique, one can either modify an observed action so that it is recognized as an instantiation of an existing EMAT category, or, if necessary, add a new EMAT category to accommodate a truly new type of action.

When creating the phrase for a new EMAT category, strive to make the phrase as general as possible but just specific enough to allow a match to occur when the observed action is read by **id**.

Protocol for identifying an action's target(s):

To determine the target of an action, ask the question: *What group would have any significant response to the action or be significantly affected by it?*

If an action targets the actor's group, the target is the same as the actor.

To simplify the IntIDs model, code only presidents as targets of reports issued by EPAs, and code only EPAs as targets of conservation NGO reports.

9.4.2   EMAT coding examples

Example 1

The first example is an article taken from the Ugandan newspaper *The Monitor* by the web news service *All Africa* entitled 'Row Erupts Between Kabarole Residents and UWA' and published August 4, 2008. The acronym 'UWA' stands for the *Uganda Wildlife Authority*. The following quote is from this article:

> The row started on July 20 when UWA officials while opening the Tooro-Semuliki Wildlife Reserve boundaries went to the three parishes of Kituule, Kijura and Kabende and cleared people's banana plantations.

This action is the 676th action in the group actions database for the East African cheetah EMT is coded as

```
676  8  4  08  All_Africa  1  UWA  1  URR
     destroy_crops_of_reserve_encroachers
     1  Uganda  1  Tooro-Semuliki_Wildlife_Reserve
     7  20  08
```

Example 2

The second example is an article taken from the online edition of the Kenyan newspaper *The National* entitled 'Lion Killings Lead to Calls for Pesticide Ban' and published July 31, 2008. The following quotes are from this article:

> In April, two male lions were killed in the Maasai Mara, a world famous game reserve. A postmortem showed the lions had been poisoned with carbofuran ...
> Conservationists say that villagers living near game parks are using the pesticide to kill predators of their livestock.

This action is the 683th action in the East African cheetah EMT database and is coded as

```
683  7  31  08  The_National  1  KRR  1  lions
     poison_lions_to_stop_livestock_predation
     1  Kenya  1  Maasai_Mara
     4  1  08
```

9.5   Actions history data for the East African cheetah EMT

9.5.1   Data sources

An actions history data set is formed by coding stories posted on the websites of the following organizations (called *news story sources*): Earthwire, Africa

Online, All Africa, Planet Ark, EnviroLink, UN Wire, Afrol, ENN, BBC News, World Bank DevNews, WildAfrica Environmental News, National Geographic News, LawAfrica, Kenya Government, Kenya Wildlife Service, Daily Nation, East-African, IndexKenya, Tanzania News, Business Times, Business News, Sunday Observer, Family Mirror, The Guardian, The Express, Tanzania Lawyers' Environmental Action Team, Uganda Government, The Monitor, The New Vision, One World, Uganda Ministry of Water, Lands and Environment, and the Uganda Parliament. The data set currently contains stories from 1997 to 2010. Links to these news story sources are collected in the `data_sets` page under the EMT example website: `www4.uwm.edu/people/haas/cheetah_emt`.

Currently, news story source websites are scanned for stories every two months. Scanning would ideally be conducted continuously in order to maximize the chance of observing action–reaction pairs. An automatic web-scanning system is being developed that will scan continuously for stories.

There is one special case, however, when the constant-time-interval sampling schedule is unbiased. This is when it can be assumed that actions occur randomly over time and the time between an action and its reaction is usually less than the scanning interval. In this case, the temporal gaps in news story coverage may not have a large effect on the performance of the fitted model. This is because the model, although fitted to an actions history that has temporal gaps, generates complete action–reaction pairs, and action–reaction–re-reaction triads. The fitting procedure, therefore, will fit a model that produces these observed action–reaction pairs and action–reaction–re-reaction triads. In other words, a model fitted to such a set of actions need not be systematically biased.

A separate file contains a list of acronyms and their expansions used in an EMAT database. This file is read by **id** whenever an EMAT database file is read. For example, the file `eastafgroups.dat` contains records such as

```
KRR Kenya rural residents
UWA Uganda Wildlife Authority
```

9.5.2   Discussion

Figure 4.1 in Chapter 4 plots the complete actions history. This figure shows that actions begin to repeat themselves after about two years of data collection – suggesting that finite repertoire decision making is being practiced and hence the group ID of a finite repertoire of actions developed in Chapter 6 is appropriate for the application to East African wildlife management.

9.6   Conclusions

Political action taxonomies were briefly reviewed. The BCOW taxonomy, in particular, was described in detail. This taxonomy was extended and modified to allow it to represent ecosystem-affecting actions. This new taxonomy is referred to as the Ecosystem Management Actions Taxonomy (EMAT). A complete coding manual was given for constructing action entries for an EMAT database.

A central premise of any taxonomy is that there is some stability or repeatability to the categories. The actions history data set for the East African cheetah EMT exhibits many repeated actions. Hence, the approach taken in this book of identifying a finite repertoire of actions for each group may be appropriate to modeling the political behavior of a set of ecosystem-affecting groups.

The two-month sampling frequency used to build the example's actions history data set is not conducive to detecting action–reaction pairs. A combination of higher sampling frequency, more in-depth news story information, and more complete coverage of political events in the habitat-hosting countries would address this deficiency. Specifically, automatic acquisition of news stories would make higher sampling frequencies easy to achieve. Such an automatic group action acquisition system would scan news stories for keywords in order to detect and codify relevant actions from news stories that are published online.

As a first step towards automating news story collecting, the **id** language contains the `parse_stories` relation. This relation is one of the relations to the `prepare_data` qualifier of the `report` statement. The relation will automatically create EMAT actions history entries for those news stories that it can successfully parse.

10

Ecosystem data

10.1 Introduction

This chapter contains a description of ecosystem variable data used in Chapter 11 to estimate the parameters of the ecosystem ID in Chapter 8. This ID has nodes to represent within-region cheetah abundance, prey abundance, and habitat suitability. Because cheetahs are elusive and exist over large areas, direct counting of a region's entire population is not done. Instead, various methods are used to estimate their abundance. Herbivores weighing less than about 60 kg are a cheetah's preferred prey, although less elusive and more numerous prey also exist over large areas making, again, direct counting of entire populations not possible. Hence, prey abundance is also estimated using some of the methods used for the cheetah.

The ecosystem ID's output nodes are functions of cheetah abundance and prey abundance. Observations on these nodes are indexed by time and region. These observations form data sets that support several different abundance estimators through the device of having abundance *cause* the abundance estimator's statistic. This causal relationship is naturally represented by an ID and is employed here to show how several estimation methods can be incorporated into one ecosystem ID.

The habitat suitability nodes are climate, protected areas, and unprotected landuse. Because these nodes represent spatial information, GIS tools are used to collect data on them.

Prey availability may be the pivotal factor in a large predator's viability (Hayward *et al.* 2007). Hence, cheetah and prey abundance estimation is discussed before that of mapping habitat suitability. Accordingly, this chapter's layout is as follows. First, some issues are discussed surrounding wildlife monitoring programs. In Section 10.3, details are given of several abundance estimation methods. Then,

Improving Natural Resource Management: Ecological and Political Models   Timothy C. Haas
© 2011 John Wiley & Sons, Ltd

data on cheetah and prey abundance for use with the East African cheetah EMT is presented in Section 10.4 along with sources of these data sets and details of the attendant collection methods. Lastly, maps of the above habitat suitability variables for the East African cheetah EMT are given.

All data sets presented in this chapter can be downloaded from www4.uwm.edu/people/haas/cheetah_emt/datasets.

10.2   Wildlife monitoring

10.2.1   Overview and issues

Issues surrounding wildlife monitoring will be delineated here by looking at the specific problem of estimating cheetah abundance in East Africa.

Bashir *et al.* (2004) review several methods for estimating cheetah abundance in a particular area (referred to therein as *census techniques*). These *sampling methods* are:

- systematic direct counting of live animals via transect surveys (see Thomas *et al.* 2002, Buckland *et al.* 2001, Silveira *et al.* 2003, or Thompson 1992, chapter 17);

- automated camera traps (see Rowcliffe *et al.* 2008);

- spoor counting (see Stephens *et al.* 2006);

- DNA analysis of scat retrieved by trained dogs (see Foran *et al.* 1997 Long *et al.* 2007);

- analysis of photographs of cheetahs taken by tourists; and

- questionnaires and/or interviews (see Gros 1998).

Other than a program for gathering tourist-taken photographs of cheetahs run by the Tanzania Wildlife Research Institute (see below), none of these methods have been implemented in an ongoing cheetah monitoring program in any country.

The ideal method for observing cheetah abundance in an area is to directly count every individual in the area. Because this method is not practical due to its high labor cost, it will not be discussed further.

The next most accurate sampling method would be to directly count animals while walking a large number of transects that have been laid out randomly or systematically over the entire area occupied by the wildlife population. Although feasible for small areas, this method is politically unrealistic for entire political districts due to its expense and the disruption to landuse activities that its implementation would cause.

The next most accurate method would be a regular grid of camera traps wherein the side of a grid cell was no more than about 2 km. This method is also politically unrealistic.

The next most feasible method would be a monitoring program that uses a mix of transect surveys, camera traps, spoor identification, DNA analysis of dog-retrieved cheetah scat, and interviews. If such a program were to be permanently funded in habitat-hosting countries, it would be close to ideal from the perspective of being able to precisely estimate the parameters of the ecosystem ID.

The political realities, however, of monitoring a terrestrial species within several countries are: (a) purposeful monitoring is usually not funded at a level that permits accurate population size estimates; and (b) if funded in the future, such a monitoring program would most likely exist for only a few years before funding was withdrawn for a variety of reasons. Such realities are particularly the case in developing countries such as Kenya, Tanzania, and Uganda.

In the face of such political realities, one approach to developing an ongoing species abundance estimation program is to gather observations that result from nonintentional, opportunistic observations that occur in several forms across space and time. The idea here is that instead of trying to determine the single best sampling method to use, an ecosystem manager needs a tool that enables him/her to optimally combine observations on a species from any available, opportunistic (passive) source so that the best use of all extant observations can be made to estimate the ecosystem ID's parameters. Funded monitoring programs should, of course, be pursued. But the EMT should not be constructed under any assumption that such funding will exist in perpetuity. For this reason, the ecosystem ID of Chapter 8 represents several different species abundance estimation methods: direct counting on transects, observing presence–absence, and camera trapping (capture–recapture data). Having such extensions to the ecosystem ID allows all opportunistically gathered observations to be used to estimate its parameters.

Modeling sampling methods within the ecosystem ID is consistent with the standard approach in wildlife sampling of deriving population estimators under assumptions about the population's behavior. For example, Karanth and Nichols (1998) estimate tiger abundance densities under several different models of how an individual animal changes its behavior after being photographed by the camera trap for the first time.

Here, because an ecosystem model has already been constructed to model cheetah population dynamics (see Chapter 8), a natural extension is to include the mechanisms of several different sampling methods into this existing model. Doing so has two benefits: first, causal relationships between the true population size (N_t) and observations are made explicit; and, second, as mentioned above, the use of multiple, disparate, and opportunistic data sources allows the manager to drive parameter estimate uncertainty down by scavenging data collected under programs that are outside the ecosystem manager's control and/or budget.

Combining different statistical estimators of a common parameter can, however, be complicated – see, for example, Handcock *et al.* (2000). But combining abundance estimators within an ID is straightforward and is accomplished through the addition of terminal nodes that represent different estimation methods. These terminal nodes have conditional distributions in N_t that are derived from the associated method, for example, camera trapping. This simple procedure for representing different abundance estimation methods is made possible because the ecosystem ID is a single, explicit stochastic model of the latent random variable N_t (cheetah abundance) that drives all observable random variables.

Data sets on any of the abundance estimation methods that contain missing values and/or unequal spatio-temporal sampling intervals require no modifications to the consistency analysis parameter estimator described in Chapter 11.

10.2.2   Direct and indirect methods

In the next section, several direct methods and one indirect method are reviewed for estimating wildlife abundance. These methods are chosen from a large literature on the estimation of wildlife abundance for their potentially low labor requirements and low levels of disruption.

Direct methods involve visually counting a subset of a region's population and then using this number with a statistical estimator to arrive at an estimate of that region's population size. Such counts are acquired by (a) performing a purposeful, funded monitoring survey using paid observers, (b) passively receiving volunteered animal sighting reports from rural residents, pastoralists, or tourists, (c) running a purposeful, funded camera trap survey, or (d) (recently) counting a subset of the area's population by analyzing very high-resolution satellite photographs. Hereafter, the phrase *nonsurvey sightings* will refer to sightings data acquired from volunteered reports, interviews, or the administration of questionnaires.

The indirect method described below involves the counting of occurrences of animal spoor (tracks, paw or hoof prints) followed by a statistical estimate of the population's size. Spoor counting surveys are typically purposeful and funded.

10.3   Wildlife abundance estimation methods

10.3.1   Direct method 1: transect surveys

A *strip transect* survey consists of a surveyor walking a straight line down the middle of a strip of area having a constant, predetermined width. All animals that are in this strip of area are counted. A *fixed radius point count* consists of the surveyor standing at the center of a circle with a predetermined radius and counting all animals within the circle (Hutto *et al.* 1986). A *line transect* survey consists of the surveyor walking a straight line and recording each animal's distance away from the surveyor measured perpendicular to the line being walked. A *point transect* survey consists of a surveyor standing at a single point and recording each animal's distance away from the surveyor.

To estimate the fixed number of animals in a region having area A, one multiplies the number of observed animals by the ratio of the region's area to the area of observation (Thompson 1992, p. 178). In effect, this estimate is first estimating animal density, and then multiplying this estimated density by the area of the region that contains the population. Strip transects and fixed radius point counts assume the same chance of detecting an animal no matter where the animal is located inside the strip or circle. Line and point transects use a detection probability as a function of the distance between the surveyor and the animal. These functions are estimated from observations on animal distances from the surveyor. The estimated function is represented by an *effective* strip (half-)width (Thomas *et al.* 2002). Let a be the actual observation area (for a strip transect or fixed radius point count) or the effective area (for a line or point transect).

Let Y_t be the number of animals observed in the strip area at time t. Design-based inference essentially views animal location as random and hence takes the expected value of Y_t to arrive at the estimate of N_t as $(A/a)y_t$ where y_t is the expected value of Y_t.

An alternate approach to this problem is to view the number of animals in the region, N_t, as random. Then N_t *causes* the number of observed animals through $Y_t = (a/A)N_t$. Because N_t and Y_t are linearly related, taking the expected value of either N_t or Y_t does not change the functional form of this relationship. Therefore, one may define a chance node Y_t that has one parent, N_t, and one parameter, a/A.

Let μ be the strip's predetermined half-width, a fixed radius point count's predetermined radius, a line transect's effective half-width, or a point transect's effective radius. Let L be the length of the strip or the length of the line transect's centerline. Then $a = 2\mu L$ for strip and line transects, and equals $\pi\mu^2$ for fixed radius point counts and point transects (Thomas *et al.* 2002).

For line and point transects, μ is estimated from data on animal distances rather than from data on animal numbers alone. A variety of estimation techniques have been developed, see Thomas *et al.* (2002).

10.3.2   Direct method 2: nonsurvey sightings

10.3.2.1   Data collection procedure

Gros (1998, 1999, 2002) uses an interview technique to conduct cheetah abundance surveys in Kenya, Uganda and Tanzania, respectively.

Gros (1998) notes a distinction between reported and actual cheetah presence, that is, the lack of a cheetah sighting within a district is not equivalent to zero cheetah abundance in that district. It is also known that survey reports undercount cheetah numbers (Gros 1998). Hence, use of interview-based survey abundance values will contribute to the fitted ecosystem model under-predicting true cheetah numbers.

10.3.2.2   Estimator

Information on the number of animals in a sighting is not always available with this type of data. One way to account for the possibility of a multiple-animal sighting is to define a parameter ρ as the number of animals per sighting. Let S_t be the number of nonsurvey sightings in a region at time point t. Then S_t is caused by N_t, the true number of animals in this region at time point t, through the relationship $S_t = N_t/\rho$.

10.3.3   Direct method 3: camera traps

10.3.3.1   Data collection procedure

A camera trap is a digital flash camera that is triggered by infrared sensors. Camera traps are unbaited or baited with a 'natural lick' (Matsubayashi *et al.* 2007, Yasuda *et al.* 2007).

10.3.3.2   Estimators

Rowcliffe *et al.* (2008) describe a way to estimate species density from camera trap photographs without identifying individual animals. Standard camera trap density estimators such as those in the CAPTURE software system (see Otis *et al.* 1978) require photograph post-processing by skilled technicians to identify individual animals by their markings. This standard approach of identifying individual animals was found to be reasonably accurate for estimating cheetah abundance in South Africa (Marnewick *et al.* 2008).

The method of Rowcliffe *et al.* (2008), however, estimates species density by explaining the number of photographed animals with a function of species density at time t (G_t), average animal group size (g), animal speed of movement (v) (also referred to as the animal's *day range*), the maximum distance at which the camera can detect an animal (r), and the angle of the camera's detection zone (θ). This detection zone is shaped like a slice of pie with the camera at the pie's center. If A is the area of the region that contains the population of size N_t, $G_t = N_t/A$.

Let Y_t be the number of photographs triggered by animals over the time interval u for purposes of estimating animal density at time t. Rowcliffe *et al.* (2008) derive

$$G_t = \frac{gY_t}{u} \frac{\pi}{vr(2+\theta)}. \tag{10.1}$$

Because the ecosystem ID of Chapter 8 is causal in animal density, Y_t is represented as a consequence of G_t in the ecosystem ID through rearrangement of Equation 10.1:

$$Y_t = \frac{G_t u}{g} \frac{vr(2+\theta)}{\pi} = \frac{uvr(2+\theta)}{\pi g A} N_t \tag{10.2}$$

with parameters v, g, r, and θ. Actually, Rowcliffe *et al.* (2008) begin their derivation with Equation 10.2 wherein N_t is replaced with $E[N_t]$. Because expected value is a linear operator, however, we may say that Equation 10.2 holds.

Rowcliffe *et al.* (2008) use this method to estimate known ungulate populations and find the estimator to be reasonably accurate. Based on simulation studies, these authors recommend at least 20 camera placements and caution that the estimator's accuracy and freedom from bias are tied to having good estimates of v and g.

10.3.4   Direct method 4: presence–absence data

10.3.4.1   Data collection procedure

An area is partitioned into a grid of cells. Each of these cells is visited and a '1' is recorded if the species is present, and a '0' otherwise. Nonsurvey sighting data can be used to indicate presence–absence by transforming the data to be the value '1' if the number of reported animals in a cell is greater than zero, and '0' otherwise. Such an approach might be pursued if a nonsurvey sightings data set is deemed to have high measurement error.

10.3.4.2   Estimators

Binomial distribution estimator Zhou and Griffiths (2007) give abundance estimators for the sampling method of surveying $C = A/a$ cells in a grid that covers a monitoring region with area A and grid cell area equal to a. These estimators only need the recording of the presence or absence of the species in each cell. Let D be the average detection probability across cells, and N_t $(= \sum_{i=1}^{C} N_{ti})$ be the true total number of animals over the entire grid of cells at time t.

Let the random variable S_{ti} be the number of surveys for which presence is detected out of m_{ti} surveys conducted on cell i. If animal aggregation behavior is ignored, then a mixture of two binomial distributions yields the joint probability mass function of $S_{t1}, \ldots, S_{tC}$:

$$P(S_{t1} = n_{t1}, \ldots, S_{tC} = n_{tC} | a, A, m_{t1}, \ldots, m_{tC}, D, N_t)$$

$$= \prod_{i=1}^{C_1} \left\{ \binom{m_{ti}}{n_{ti}} D^{n_{ti}} (1 - D)^{m_{ti} - n_{ti}} \left[1 - \left(1 - \frac{a}{A} \right)^{N_t} \right] \right\}$$

$$\times \prod_{i=1}^{C_0} \left\{ (1 - D)^{m_{ti}} \left[1 - \left(1 - \frac{a}{A} \right)^{N_t} \right] + \left(1 - \frac{a}{A} \right)^{N_t} \right\} \qquad (10.3)$$

where n_{ti} is the observed number of surveys for which the species is detected in cell i at time t out of the total number of surveys conducted on that cell, m_{ti}. The value of C_1 is the total number of sampled cells for which $n_{ti} > 0$, and C_0 is the total number of sampled cells for which $n_{ti} = 0$. If only one survey is conducted, this equation simplifies to a multivariate Bernoulli distribution:

$$P(S_{t1} = n_{t1}, \ldots, S_{tC} = n_{tC} | a, A, D, N_t)$$

$$= D^{C_1} \left[1 - \left(1 - \frac{a}{A} \right)^{N_t} \right]^{C_1} \times \left\{ (1 - D) \left[1 - \left(1 - \frac{a}{A} \right)^{N_t} \right] + \left(1 - \frac{a}{A} \right)^{N_t} \right\}^{C_0}$$

$$(10.4)$$

with the single parameter D.

Areal detection fraction estimator Another way to use presence–absence data to estimate abundance is with the *areal detection fraction* first described in Haas (2001). In the ecosystem ID, the observable random variable of this estimator is the 'Cheetah Detection Fraction' node (D_t^C). This node's value is the fraction of a region over which animals have been detected.

The node D_t^C is caused by N_t through

$$D_t^C = \frac{N_t/A - \xi^C}{\rho^C - \xi^C} I_{\{\xi^C < N_t/A < \rho^C\}}(N_t) + I_{\{\rho^C < N_t/A\}}(N_t) \qquad (10.5)$$

with parameters ξ^C and ρ^C. The parameter ξ^C is the largest density below which humans in the region cannot detect the animal, and ρ^C is the smallest density for which humans will always detect the animal.

Abundance observations are converted to areal detection fractions by dividing the area over which such observations are nonzero by the total area of the region.

10.3.5 Direct method 5: hyper-spatial resolution satellite imagery

10.3.5.1 Data collection procedure

Recent launches of satellites that can take very high-resolution photographs has made it possible to detect animals from space. Currently, this technology can accurately detect animals in low-vegetation habitats when their reflectance (brightness) is high. For example, using QuickBird-2 images, Sasamal *et al.* (2008) detect groups of flamingos, and Barber-Meyer *et al.* (2007) detect groups of penguins. Lucieer (2010) detects groups of elephant seals in a WorldView-1 image. And individual elephants in Kenya's Amboseli National Park can be seen in a QuickBird-2 image (Ramanujan 2010). QuickBird-2 has a 60 cm resolution (QuickBird-1 was lost on launch), and WorldView-1 has a 50 cm resolution. WorldView-2 has a 46 cm resolution and GeoEye-1 has a 41 cm resolution; but images from either of these two satellites can only be purchased at a 50 cm resolution due to US regulations, see eMap-International (2010).

Critical aspects of this method are (a) its expense can be expected to fall, and (b) gathering such images can be accomplished without the expense of field expeditions.

10.3.5.2 Estimator

Once images are processed with (say) the animal-counting algorithm of Laliberte and Ripple (2003), standard line transect estimators can be used to estimate abundance if only a portion of a region has been imaged. If, on the other hand, the entire region has been imaged, then N_t is simply the number of animals counted in the image.

10.3.6   Indirect method: spoor counts

10.3.6.1   Collection method

Observers walk linear transects and count the number of animal tracks (spoors) that cross the transect line.

10.3.6.2   Estimator

Knowing the animal's average speed allows an estimate of the abundance to be computed via the Formozov–Mayshev–Pereleshin formula (see equation (5) in Stephens *et al.* 2006). This formula can be derived from the Rowcliffe *et al.* (2008) formula by setting θ to 0. Hence, the ID represents spoor count observations on these transects with

$$Y_t = \frac{G_t u}{g} \frac{2vr}{\pi} = \frac{u2vr}{g\pi A} N_t \tag{10.6}$$

with parameters v, g, and r.

10.3.7   Summary

For ease of comparison with the ecosystem ID of Chapter 8, all of these estimators are summarized in Table 10.1.

Table 10.1   Summary of wildlife abundance estimators.

Method	Observed variable(s)	Causal form of estimator	Parameters
Transects	Y_t	$Y_t = (a/A)N_t$	μ
Nonsurvey sightings	S_t	$S_t = N_t/\rho$	ρ
Camera traps	Y_t	$Y_t = \dfrac{uvr(2+\theta)}{\pi g A} N_t$	u, v, g, r, θ
Presence–absence	$S_{t1}, \ldots, S_{tC}$	Distributed as mixture-binomial	D
Areal detection fraction	$D_t^C = C_1/C$	$D_t^C = \dfrac{N_t/A - \xi^C}{\rho^C - \xi^C} I_{\{\xi^C < N_t/A < \rho^C\}}(N_t)$ $+ I_{\{\rho^C < N_t/A\}}(N_t)$	ξ^C, ρ^C
Hyper-spatial resolution imagery	Y_t	Same as strip transects	μ
Spoor counts	Y_t	Same as camera traps with $\theta = 0$	u, v, g, r

10.4   East African cheetah and prey abundance data

10.4.1   Cheetah data

10.4.1.1   Transect surveys

Over the years 1999 to 2001, Maddox (2003) executed line, strip, and point transect surveys on several species in three regions of Tanzania. Because the observed cheetah numbers are small, this author does not estimate μ and instead, for purposes of estimating cheetah density, uses a constant value of 200 m. The cheetah data from these surveys (Table 10.2) is constructed here by assuming an observed *group* of cheetahs always contains exactly two individuals.

10.4.1.2   Nonsurvey sightings

Gros (1998, 1999, 2002) presents nonsurvey sightings of cheetahs. These sighting reports are used here to compute areal detection fraction values (see below).

Recently, the Tanzania Wildlife Research Institute has been posting on the Web a map of cheetah sightings across Tanzania (TMAP 2008). The cheetah sighting locations shown in Figure 10.1 were digitized from such an online map with **id**'s GIS tools. All of these cheetah sightings are collected in Table 10.3.

Evans (2004) reports 31 nonsurvey sightings in Kenya's Nakuru Wildlife Conservancy between the years 2000 and 2002.

10.4.1.3   Areal detection fraction

Table 10.4 contains such a data set for Kenya. This data is computed from entries in Table 6 of Haas (2001).

10.4.2   Prey data

10.4.2.1   Aerial surveys

Prey abundance values for 1977–1985 in Kenya are given in Mbugua (1986) and Peden (1984). This data was collected by directly counting animals from a

Table 10.2 Transect survey sightings reported by Maddox (2003). Numbers in parentheses are total transect length in kilometers. It is assumed here that only one point transect was taken per region.

Region	Transect type		
	Line	Strip	Point
Loliondo	14 (405)	3 (10)	24
Ngorongoro	11 (375)	0 (28)	0
Serengeti	16 (1092)	0 (5)	51

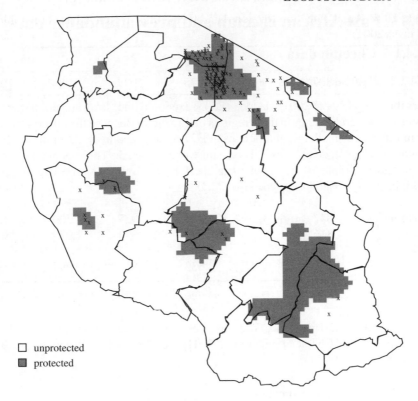

Figure 10.1   Cheetah sightings reported by tourists in Tanzania over the time period April 1, 2003 to December 1, 2008. Locations are plotted over the map of Tanzania's districts and protected areas.

Table 10.3 Nonsurvey sightings from tourists in Tanzania over the time period April 1, 2003 to December 1, 2008.

District	Count
Mara	30
Rukwa	9
Shinyanga	33
Tabora	1
Mbeya	2
Singida	3
Lindi	2
Kilimanjaro	1
Iringa	2
Dodoma	2
Arusha	45

Table 10.4   Areal detection fraction data on cheetahs in Kenya based on Table 6 of Haas (2001).

Region	D_t^C					
	1887	1962	1975	1977	1986	1990
Marsabit	1	1	1	1	1	1
Eastern	1	1	1	1	1	1
Samburu	1	1	.327	1	1	1
Tsavo	1	1	1	1	1	1
Masailand	1	1	1	1	1	1
Laikipia	1	1	1	1	1	1
Nakuru	1	0	0	1	1	1
Western	.847	.325	.246	.927	.300	.073
Central	.866	.827	.745	.841	.938	.856
Turkana	1	1	0	1	1	1
Coastal	1	1	.633	.808	.994	1

low-flying aircraft. This data is used to compute areal detection fraction values, D_t^H as defined in Chapter 8, Section 2.5. The results of these computations are shown in Table 10.5.

10.4.2.2   Line transect surveys

Table 10.6 contains line transect survey counts of individuals taken by Maddox (2003) on seven prey species. The table's computed or assumed values of μ are based

Table 10.5   Areal detection fraction observations computed from herbivore abundance data reported by Mbugua (1986) and Peden (1984).

Region	D_t^H						
	1977	1978	1980	1981	1982	1983	1985
Marsabit	1	1	—	1	—	—	.928
Eastern	.606	.837	.121	—	—	.147	.202
Samburu	1	.696	.013	.130	.024	—	.505
Tsavo	—	—	—	—	—	—	—
Masailand	1	1	1	.965	1	1	—
Laikipia	1	1	—	1	1	—	1
Nakuru	—	—	—	—	—	—	—
Western	—	—	—	—	—	—	—
Central	.075	.096	.096	—	—	.042	—
Turkana	.671	—	—	.370	.456	—	—

Table 10.6 Line transect survey prey counts, estimated values of survey half-widths (μ), total transect length (L), and region areas (A) as reported by Maddox (2003, Tables 7, 10, 98, and pp. 52–54). Total prey abundance is computed using these counts and values on constants with the line transect abundance estimator of Section 10.3.1.

Prey species	Loliondo	Ngorongoro	Serengeti
Dik-dik	53	22	0
μ (m)	200	200	200
Grant's gazelle	6209	10393	8170
μ (m)	288.1	315.63	286.41
Impala	1604	75	0
μ (m)	200	200	200
Kongoni	956	20	1715
μ (m)	417.28	270.39	357.96
Steinbuck	25	24	0
μ (m)	200	200	200
Thomson's gazelle	5068	9110	4381
μ (m)	174.16	139.23	164.28
Topi	330	0	881
μ (m)	184.34	200	432.88
L (km)	405	375	1092
A (km^2)	6734	8292	2200
B_t	412294	993732	47932

on distance data collected on these transects. Total prey abundance is computed using the table's values for A, μ, and L.

The Loliondo and Serengeti reserves are in Tanzania's Mara district while the Ngorongoro reserve is in the Arusha district. In Chapter 11, the parameters A, μ, and L are fixed at the values given in Table 10.6 during the consistency analysis

Table 10.7 Nighttime prey sightings reported by Wambua (2008) in the Kenyan districts of Machakos and Makueni.

Prey	Count
Cape hare	217
Dik-dik	46
Duiker	100
Grant's gazelle	30
Impala	24
Reedbuck	15
Steinbuck	82
Thomson's gazelle	61

parameter estimation computation. Hence total prey values appear as observations on B_t for the year 2001 in that analysis.

10.4.2.3   Nonsurvey sightings

Wambua (2008) reports prey sightings outside the towns of Machakos and Makueni in Kenya (Table 10.7). These towns are located in the ecosystem ID's Central region of Kenya. Only the night sightings are used here.

10.5   Data on cheetah habitat suitability nodes

Chapter 8 contains a description of how the habitat nodes of climate (C), proportion of land in protected areas (R_t), and unprotected landuse (L) influence cheetah and prey populations.   Figures 10.2–10.4 give observations on these nodes across the three countries of the East African cheetah EMT.

The association between these habitat suitability indices and cheetah abundance may not be strong. For example, Bashir *et al.* (2004) note that habitat suitability predictions of cheetah abundance are poorly correlated with actual cheetah counts because cheetahs exist over a wide spectrum of habitat type and are not territorial.

10.5.1   Map sources

10.5.1.1   Software for extracting data from maps

Reference is made below to **id**'s GIS tools. These tools, documented in Chapter 5, include an on-screen digitizing tool, a surface digitizing tool, and a georeferencing tool.

10.5.1.2   Boundaries: information sources

All digitization reported below was performed with **id**'s GIS tools.

Kenya's region boundaries were digitized from a figure in Gros (1998). Several of Tanzania's administrative district boundaries were digitized from the Reise map of Tanzania (Reise 2004). The rest were digitized from the United Nations PDF map of Tanzania (UN 2008a) after converting the PDF file to a GIF file with Adobe Photoshop. Uganda's administrative district boundaries were digitized from the United Nations PDF map of Uganda, see UN (2008b).

Climate regions were found by digitizing selected surfaces from the Food and Agriculture Organization (FAO) Landcover maps of Kenya, Tanzania, and Uganda, see FAO (2008). These surfaces were then grouped to produce the levels shown in Figure 10.2.

The UN PDF maps of Kenya and Tanzania were used to digitize Kenya's and Tanzania's protected areas, respectively (see UN 2008a). Uganda's protected

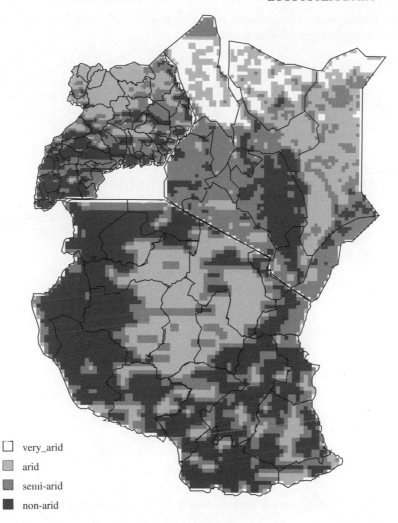

□ very_arid

  arid

  semi-arid

■ non-arid

Figure 10.2   Climate zones of Kenya, Tanzania, and Uganda.

areas were digitized from Uyaphi's Uganda Search Map, see Uyaphi (2008). These protected areas are displayed in Figure 10.3.

Unprotected landuse regions were found by digitizing selected surfaces from the FAO Farming Systems maps of Kenya, Tanzania, and Uganda, see FAO (2008). These surfaces were then grouped to produce the levels shown in Figure 10.4.

10.5.1.3   Data extraction from copyrighted maps

The extraction of data from a database, for example, digitization of a political district's boundary from a map (the database), is exempt from US copyright law,

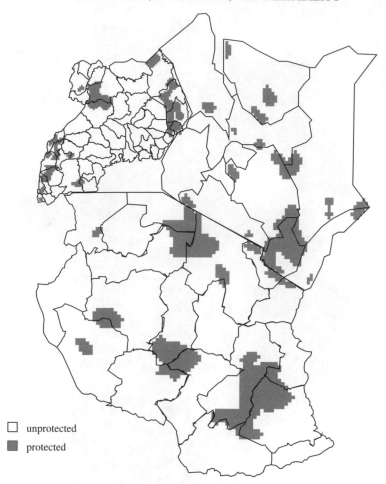

Figure 10.3   Protected areas maintained by the governments of Kenya, Tanzania, and Uganda.

see for instance BitLaw (2009a, 2009b). This exemption is because entries in a database are viewed as *basic facts* which are, by US copyright law, not protectable by US copyright. In other words, the presentation of a database is copyright protected but the entries themselves are not.

Therefore, the use of **id**'s GIS tools to extract spatial information from copyrighted maps is a legal activity with respect to US copyright law if the following entities are basic facts: (a) administrative district boundaries, (b) boundaries of homogeneous climate regimes, (c) boundaries of state-controlled protected areas, and (d) boundaries of typical landuse. The position taken in this book is that these entities are indeed basic facts and hence no US copyright infringement occurs when an analyst uses **id**'s GIS tools to digitize administrative regions such as

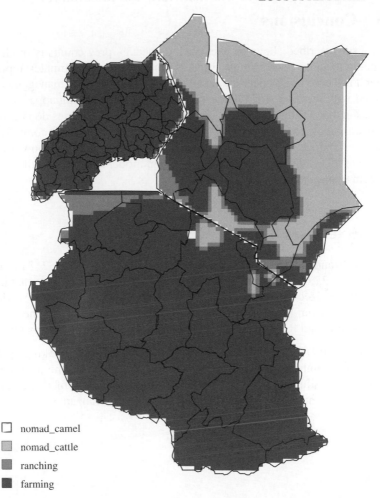

□ nomad_camel

▨ nomad_cattle

■ ranching

■ farming

Figure 10.4   Unprotected area landuse regions of Kenya, Tanzania, and Uganda.

those in Chapter 8, and the climate, protected areas, and landuse regions presented above. It is believed that, to challenge this legal position, the makers of a copyrighted map would need to exhibit a legal document that states that they hold the copyright on the set of latitude–longitude coordinates that defines a sovereign country's border.

To avoid US copyright infringement when performing such digitization, an analyst should restrict him/herself to digitizing only boundaries as expressed in latitude–longitude coordinates (not the map maker's chosen map transformation) and avoid digitizing any other features of the map such as details of how such basic facts are presented. Examples of presentation details include color selections and artistic renderings, for example, the symbol used to designate mountains.

10.6   Conclusions

The data set described above on cheetah and cheetah–prey counts is small, which contributes to large uncertainties in ecosystem parameter estimates reported in Chapter 11. These large uncertainties reduce the reliability of management plans computed with the East African cheetah EMT. Lack of funding appears to be the main reason that wildlife conservation agencies in these countries do not run more comprehensive monitoring programs. The data sets in this chapter come from several sources and sampling methods. This, then, is an example of how an ID can make use of all available data to fit an ecosystem model for purposes of supporting ecosystem management decision making.

10.7   Exercises

1. The Brandenburg Gate in Berlin, Germany, is located at $52\,°31'$N latitude and $13\,°22.6'$E longitude. There is a statute on top of this gate that includes four life-sized horses. Download an appropriate WorldView-1 or WorldView-2 image from (say) http://emap-int.com and, using the method of Laliberte and Ripple (2003), count the number of horses in this image.

2. Say that there are 2000 cheetahs in all of Tanzania. After looking up the area of this country, find all values of ξ^C and ρ^C for which the areal detection fraction estimator gives this value exactly. What sets of these parameter values are not to be expected of rural residents in Tanzania?

11

Statistical fitting of the political–ecological system simulator

11.1 Introduction

As discussed in Chapter 1, the first step towards establishing predictive validity is to statistically fit the political–ecological system simulator to data. To this end, consistency analysis (see Chapter 4, Section 4.3) is used in this chapter to fit the simulator to data. The basic idea of consistency analysis is to find a set of parameter values that cause model output on observed variables to have probability distributions that are similar to distributions derived exclusively from data, while at the same time doing minimal violence to a set of parameter values (the hypothesis values) that cause the model to behave in ways that are consistent with political and ecological theory.

This chapter is organized as follows. First, algorithms are given for initializing consistency analysis when the data set contains an actions history. Then, consistency analysis is used to fit the East African cheetah EMT simulator to a data set formed by combining the political actions data of Chapter 9 and the ecosystem data of Chapter 10. This combined data set is referred to as a *political–ecological data set*.

Improving Natural Resource Management: Ecological and Political Models   Timothy C. Haas
© 2011 John Wiley & Sons, Ltd

11.2   Consistency analysis applied to an actions history

11.2.1   Matching IntIDs output to observed actions

Refer to Chapter 4, Section 4.3, for the mathematical notation used to describe consistency analysis and for definitions of its agreement measures. Refer to Chapter 6, Section 6.3, for definitions of group ID in-combinations, out-combinations, and in–out pairs.

Consistency analysis is used to fit a model to an observed actions history under the assumption that an observed group decision is a function of the true values of parameters of that group's ID. In particular, an observed decision is not viewed as a realization from a random generator of action–target combinations, but rather as the value of a function of the group ID's true parameter values. The rationale is that since all chance nodes in a group ID are latent, there is little justification for any assumption other than that the group is, in effect, performing an exact computation with the true ID parameter values of the marginal distribution of the utility node under each in–out pair being entertained, and selecting the out-combination that gives the largest value of expected utility given the in-combination. This rationale leads to the assumption that observed actions are outputs of a deterministic function of the in-combination and the group ID's true parameter values. Under this assumption, matching as many out-combinations as possible in an actions history data set needs to be the first priority of a statistical estimation method.

As discussed in Chapter 9, a two-month sampling interval is used to collect actions history data. Because of this sampling protocol, not all actions are observed. Hence, there is some variability in the final fitted model that is due to the use of a sample rather than the complete actions history during the estimation procedure. This source of variability in parameter estimates is typically referred to as *sampling variability*. An additional aspect of an observed actions history is that in-combinations are not observable.

There is another source of variability in the parameter estimates: model misspecification. The rudimentary models of group decision making described in Chapters 6 and 7 do not capture all of the complexity of how a group such as Kenya's rural residents reaches a decision. If a model has a noise or residual term, for example the additive Gaussian noise term in a statistical regression model, then such a term can be used to represent variability due to model misspecification. The group IDs of Chapter 7 have no such term. A topic for future work is to specify and fit such a term.

When comparing an observed actions history to an ID-generated actions history, say that a *match* has occurred if, at the time point of an observed out-combination from a particular group, the associated group ID in the IntIDs model generates the same out-combination.

As mentioned above, missing actions in an observed actions history data set notwithstanding, a model that purports to explain a set of observed actions should match as many of them as possible – no matter what prior beliefs an analyst may have about how a group reaches a decision. Therefore, in the **Maximize** step of consistency analysis, $g_{CA}(\boldsymbol{\beta})$ is maximized with an optimization algorithm designed for smooth objective functions but constrained to move to a new point (value of $\boldsymbol{\beta}$) only if the match fraction at that new point is not reduced.

In other words, the algorithm that maximizes $g_{CA}(\boldsymbol{\beta})$ rejects any move to a new point if the match fraction at that point is smaller than the match fraction found in the **Initialize** step. This constraint is not equivalent to forcing c_H to be zero. Rather, the constraint implies an opportunistic approach to agreement with the hypothesis distribution: make the consistent distribution as close to the hypothesis distribution as is possible without damaging the model's initial agreement with observed actions.

An ID of a group's decision making is a function that maps in-combinations to out-combinations. A function's defining characteristic is that each unique input (here, an in-combination) cannot map to more than one unique output (here, an out-combination). Therefore, for a given actions history data set, there is an upper limit on how well a function-based model can fit such data.

11.2.2   The Initialize step

11.2.2.1   Definitions and overview

Let $n_I^{(i)}$ be the number of in-combinations represented in the ith group's ID. Let $n_O^{(i)}$ be the number of out-combinations represented in the ith group's ID. Let m be the number of group IDs in the simulator's IntIDs model. Call the set of all in–out pairs that an ID can produce an *in–out pattern*.

An in-combination that causes no out-combination is said here to cause the $(n_O^{(i)} + 1)$th out-combination, called the *null* out-combination. Therefore, for the ith group's ID, there are $(n_O^{(i)} + 1)^{n_I^{(i)}}$ possible in–out patterns because each in-combination maps to exactly one of the $n_O^{(i)} + 1$ out-combinations. An IntIDs model consists of a set of m in–out patterns. Call a particular set of such patterns an *in–out pattern collection*. Consequently, there are $\prod_{i=1}^{m}(n_O^{(i)} + 1)^{n_I^{(i)}}$ different in–out pattern collections. Some of these collections will match an observed actions history better than others.

Let the *match fraction* be the ratio of the number of such matched out-combinations to the total number of observed out-combinations.

Let $n_{obs}^{(i)}$ be the number of observed out-combinations from group i and recall from Chapter 4 that $g_S^{(i)}$ is the number of matched group i out-combinations. Let $f^{(i)} = g_S^{(i)}/n_{obs}^{(i)}$ be the match fraction for group i. Let $f = \sum_{i=1}^{m} g_S^{(i)} / \sum_{i=1}^{m} n_{obs}^{(i)}$ be the *overall match fraction*.

Overview   When consistency analysis is used to fit IDs to an observed actions history, its **Initialize** step (see Chapter 4) finds a point in the parameter space for which group ID-generated out-combinations match as many observed out-combinations as possible. The **Initialize** step accomplishes this by first finding the best-fitting in–out pattern collection, and then finding parameter values that cause all group IDs to generate an in–out pattern collection that is as close as possible to this initial in–out pattern collection. Finding the best-fitting in–out pattern collection is referred to here as *collection initialization* and constitutes a type of pre-conditioning computation before the **Initialize** step that is only needed when fitting IDs to an observed actions history. Call the in–out pattern collection produced by collection initialization the *initial collection*.

After an initial collection is found, the **Initialize** step executes by modifying entries in $\boldsymbol{\beta}_H$ until the IntIDs model computes an in–out pattern collection that is as close as possible to the initial collection. Call this modified set of parameter values, $\boldsymbol{\beta}_I$ (*I* for 'Initial').

Let $n_P^{(i)}$ be the number of in–out pairs in group i's initial in–out pattern that group i's ID is able to reproduce after execution of the **Initialize** step. During collection initialization, each group's in–out pattern involved reassignment of out-combinations without regard to whether the group ID would actually be able to compute these new out-combinations. Therefore, during the **Initialize** step, a set of parameter values that would allow the ID to generate all in–out pairs in the initial in–out pattern may not be found. Such failures are due to the **Initialize** step's parameter modification algorithm only being able to set parameter values so that an ID can generate up to three unique out-combinations.

11.2.2.2   Characteristics of collection initialization

A discrete search algorithm, namely, integer programming, is used to find an in–out pattern collection that maximizes the match fraction. Note that the objective function of this search is inexpensive because it can be expressed as the summation of binary-valued vectors rather than the computation of ID solutions.

Within the initial collection, let $n_{iop}^{(i)}$ be the number of in–out pairs in group i's in–out pattern. Note that this in–out pattern is built by running the IntIDs model and having group IDs react to out-combinations that are posted to the bulletin board (see Chapter 2). Let $n_{uoc}^{(i)}$ be the number of unique out-combinations in group i's in–out pattern.

Group IDs compute out-combinations as reactions to these posted out-combinations from other group IDs. Some of these in–out pairs may have their out-combination modified during execution of collection initialization to improve the match fraction, but the in–out pattern is not constructed from the data – it is constructed as the model runs. Hence, $n_{iop}^{(i)}$ has no simple functional relationship to $n_{match}^{(i)}$.

It may be that in group i's in–out pattern many in-combinations are paired to the same out-combination. Hence, a single out-combination may be repeated across many in–out pairs contained in the in–out pattern.

11.2.2.3   Justification of collection initialization algorithms

Two questions are begged by the use of discrete search for collection initialization. First, why not use exhaustive search of all possible in–out pattern collections to find the collection that maximizes the overall match fraction? The answer is that such an exhaustive search is intractable. For example, if $n_I^{(i)} = 10$ and $n_O^{(i)} = 9$ for $i = 1, \ldots, m$, then the size of the solution space is 10^{10m}. In the East African cheetah management simulator, $n_I^{(i)}$ and $n_O^{(i)}$ are at least 10 for all groups and $m = 13$, making exhaustive search of this space impractical. To see why this is, say that the objective function requires 10^{-6} seconds to compute. Then an exhaustive search of this space would require

$$\frac{10^{130-6} \text{ seconds}}{3.1536 \times 10^9 \text{ seconds/century}} = 3.188 \times 10^{114}$$

centuries of computer time.

The second question is: why not employ a search algorithm designed for smooth objective functions to find parameter values that maximize the overall match fraction, thereby combining **Initialize** steps 1.1 and 1.2 into one computation that maximizes the overall match fraction wherein β_H is used as the initial solution? The answer is that optimization algorithms designed for smooth objective functions can have difficulty in finding points that increase the overall match fraction because that function has a constant value over local regions of the parameter space. With actions history data, however, maximization of the overall match fraction is of the highest priority (see above). Therefore, to assure that this maximum overall match fraction is found, collection initialization uses integer programming methods to find an in–out pattern collection that maximizes the overall match fraction.

11.2.2.4   Collection initialization algorithms

Each group's in–out pattern needs to be modified so that the IntIDs model matches as many observed actions as possible. Two algorithms that do this are given. Running either algorithm produces an in–out pattern collection for the IntIDs model such that when a model possessing this in–out pattern collection is run over the same time period as the observed actions history, the overall match fraction value is maximal. The first algorithm, greedy search, is given next, and the second one, simulated annealing, is given in the conclusions section as it has yet to be implemented in **id**.

11.2.2.5   Greedy search

Commencing at the first observed time point and proceeding sequentially through to the last time point, begin with the first group ID and, proceeding through to the last group ID, check if the observed out-combination for that time point and group is matched by the out-combination computed by that group's ID. If they do not match, replace the out-combination in that ID's in–out pattern with the one that was observed. Then, run the IntIDs model over the entire observed time period and compute the overall match fraction. Reject this change if this trial overall match fraction is unchanged or has decreased.

Note that, at each time point, the algorithm is trying to adjust each group's in–out pattern so that the out-combination computed by that group ID in response to an IntIDs model-generated in-combination is the same as the observed out-combination at that time point. This dependence on model-generated in-combinations is essential because, as mentioned above, in-combinations are not observable.

11.2.3   Initialize step algorithm

As mentioned above, collection initialization produces an initial in–out pattern for each group. Because group IDs are functions, these patterns are constrained so that each in-combination maps to a single out-combination.

Before describing the algorithm used by the **Initialize** step to find initial ID parameter values when fitting to an observed actions history, a trivial case is noted. It may be that all in-combinations map to the same out-combination. In this case, it is only necessary to make that single out-combination the most preferred under any in-combination.

Several heuristics are employed in the **Initialize** step algorithm. They are:

1. Consider all in-combinations that map to the same out-combination as a cluster and set all their parameters to the same values.

2. Change only those conditional distributions that are conditioned on in-combinations or out-combinations that are members of the initial collection.

3. For a node with a nuisance second parent, use the conditional distribution dictated by the first parent's value regardless of this second parent's value.

4. Let nodes directly affected by an input action node or an output action node be called *direct-effect nodes*. Set the conditional distributions on nodes that are in between direct-effect nodes and a goal node so that values of direct-effect nodes are passed down the hierarchy with very little noise.

11.2.3.1   Evaluation dimensions

Sets of nodes that are influenced by input action, actor, output action, or the chosen target represent *evaluation dimensions*. These dimensions are the main criteria along which a group evaluates an in-combination or an out-combination (see Table 11.1).

Conditional distributions on Situation subID nodes represent the group's interpretation of an in-combination along these evaluation dimensions. Because these are Situation nodes, their settings are functions of in-combinations alone, that is, these settings do not change across different out-combinations under the same in-combination.

11.2.3.2   Truth table

The truth table (Table 11.2) allows an ID to represent three unique out-combinations, each a reaction to an in-combination that is a member of that out-combination's cluster of in-combinations. Analysis of the East African actions history data

Table 11.1 Nodes that represent evaluation dimensions.

Group	Dimension 1	Dimension 2
President	Economic change	Militaristic change
EPA	Economic change	Poaching action
	President change	Rural resident poaching rate
		Pastoralist poaching rate
Rural resident	Economic change	Imminent interaction with police
Pastoralist	Territory size	Imminent interaction with police
	Livestock change	
Conservation	Wildlife prevalence	President change (influences
NGO	Change (influences	maintain relations goal)
	Economic change)	

(see Chapter 9) suggests that in–out patterns with more than four unique out-combinations are rare.

Conditional distributions on Scenario goal nodes are assigned based on entries in this truth table. This table is read as follows. First, consider a 'change' node with five values, for example, *large decrease, small decrease, no change, small increase, large increase*. Let b be a conditional distribution of this node that has most of its probability mass on the node's first value (b for *bad*). Let m be a distribution that has most of its probability mass on the third value (m for *medium*). Let g be a distribution that has most of its probability mass on the fifth value (g for *good*). As shown in the following example, a group ID can produce three unique in–out pairs with these three types of conditional distributions,

Table 11.2 A truth table to relate `Situation Change` and `Scenario Change` node value combinations to values of a `Scenario Goal` node.

Situation change node	Scenario change node	Scenario goal node
b	g	g
g	b	b
b	b	b
g	g	m
g	m	b
b	m	m
m	b	b
m	g	m
m	m	m

Table 11.3 Situation subID to Scenario subID relationships that produce three unique out-combinations. Bold entries indicate desired (maximum utility) in–out pairs.

in-comb	out-comb	ECONCHG	MILRCHG	SECONCHG	SMILRCHG	SEG	SMG
1	**1**	**g**	**b**	**b**	**g**	**b**	**g**
1	2	g	b	g	b	m	b
1	3	g	b	m	m	b	m
2	1	b	g	b	g	b	m
2	**2**	**b**	**g**	**g**	**b**	**g**	**b**
2	3	b	g	m	m	m	b
3	1	m	m	b	g	b	m
3	2	m	m	g	b	m	b
3	**3**	**m**	**m**	**m**	**m**	**m**	**m**

Example

Consider a president ID. Let the three unique out-combinations and their associated in-combination clusters be numbered 1, 2, 3. For example, unique out-combination #2 is composed of 'in-comb' cluster #2 and 'out-comb' # 2. These combinations are used to form the needed conditional distribution assignments on Situation Change, Scenario Change, and Scenario Goal nodes as listed in Table 11.3. Note that, in this table, there are only three possible goal node patterns that give equally large expected values of the OGA node: (g,b), (b,g), and (m,m).

Actually, Table 11.2 was discovered by creating Table 11.3 via the following two-step procedure:

1. Propose the following plausible mappings of Situation Change and Scenario Change node values to a Scenario Goal node value. Call these Sit-Scen⇒ScenGoal mappings.

   (a) Decision makes a positive change: b,g⇒g.

   (b) Decision makes a negative change: g,b⇒b.

   (c) Decision does not improve a bad situation: b,b⇒b.

   (d) Decision makes a bad situation worse: g,m⇒b.

   (e) Decision makes a bad situation better: b,m⇒m.

   These relationships give the first two optimal out-combinations in Table 11.3 since they are mirror images of each other.

   The third optimal out-combination has to have a different goal node pattern: (m,m). But the Sit-Scen⇒ScenGoal mappings for these in–out pairs are all different than those for the first two optimal out-combinations. These mappings are:

(a) Decision does not change a tolerable situation: m,m⇒m.

(b) Decision makes a tolerable situation worse: m,b⇒b.

(c) A medium situation is made better but the decision maker does not acknowledge this fact: m,g⇒m. This relationship is admittedly hard to justify but is needed to ensure that all in–out pairs having the third in-combination result in (b,m) or (m,b) on the `Scenario Goal` nodes except for the third one (which produces (m,m)).

2. Once Table 11.3 is constructed, the unique `Sit-Scen`⇒`ScenGoal` mappings are extracted from it to form Table 11.2.

For Table 11.3 to produce a desired pattern of in–out pairs, scenario goal patterns of (b,g), (g,b), and (m,m) need to produce the same expected value on the OGA node. Further, this expected value needs to be larger than that computed under the goal patterns (m,b), and (b,m). Otherwise, for some desired in–out pair, the ID will not select the out-combination that is associated with that in–out pair's in-combination.

The entries in Table 11.3 can be used to assign conditional distributions for either two or three unique in–out pairs without modification. Given a president-type ID, one way these three in–out pairs could be characterized is as follows:

- The first in–out pair can be interpreted as the president addressing a bad military problem with an effective military solution.

- The second in–out pair can be interpreted as the president addressing a bad economic problem with an effective economic solution.

- The third in–out pair can be interpreted as the president addressing a bad combined economic and military problem with a balanced and effective economic–militaristic solution.

11.3   Consistency analysis of the East African cheetah EMT simulator

11.3.1   Optimization problem configuration

The time points at which the IDs read the bulletin board are aligned with those in the political–ecological data set of Chapters 9 and 10. Doing so allows data-based causal chains of action and reaction to be learned by the model during parameter fitting. The initializing actions are Tanzanian rural residents indirectly damaging wildlife habitat, and the Kenyan EPA increasing anti-poaching enforcement.

To reflect the mandate of Section 11.2.1 to match as many actions as possible (no matter how poorly the fitted model might agree with theories of political decision making), c_H is set to a small value, namely 0.01. As discussed in Chapter 4, Section 4.3.1, setting c_H to a small value nearly turns consistency analysis into a minimum Hellinger distance statistical estimator.

Optimization is performed separately on each group ID. For each of these group IDs, only parameters defining scenario goal nodes are modified during the **Maximize** step. A separate **Maximize** step is executed on the ecosystem ID. Each of these optimization problems consists of about 40 parameters being adjusted in an effort to maximize the consistency analysis objective function. A constrained version of the Hooke and Jeeves optimization algorithm is used to search for a point that maximizes $g_{CA}(\boldsymbol{\beta})$. See Appendix B for how this algorithm is modified to handle a stochastic objective function and for several ways to implement it on a cluster computer.

11.3.2   Consistency analysis results

11.3.2.1   Collection initialization

The greedy search algorithm described above is used to maximize $g_S^{(i)}(\boldsymbol{\beta})$ across all group IDs. This maximization procedure results in an initial collection summarized in Table 11.4. Out of the 252 action–target observations, 60 (23.8%) are matched by the initial collection. In addition, the overall action match fraction is 30.1%, and the overall target match fraction is 71.8%.

11.3.2.2   Possible ways to improve initial collection match statistics

To increase the match fraction achieved by the initial collection, the number of different out-combinations generated by an ID at each time step needs to be increased

Table 11.4 Initial collection match statistics. See Section 11.2.2.1 for symbol definitions.

Group	$n_{obs}^{(i)}$	$g_S^{(i)}$	$f^{(i)}$	$n_{iop}^{(i)}$	$n_{uoc}^{(i)}$
kenpres	23	2	0.086	8	2
kenepa	41	9	0.219	14	6
kenrr	30	11	0.366	24	4
kenpas	5	4	0.800	11	2
tanpres	5	1	0.200	7	2
tanepa	17	3	0.176	8	3
tanrr	7	3	0.428	15	5
tanpas	0	0	0.000	13	1
ugapres	13	4	0.307	9	4
ugaepa	37	5	0.135	19	6
ugarr	21	7	0.333	23	5
ugapas	6	5	0.833	11	2
ngo	47	6	0.127	25	6

so that there are more different in-combinations at each subsequent time step to support replacement of wrong out-combinations with observed out-combinations. Doing so would allow the model to fit actions history data that appears to have been generated by interacting relations rather than interacting functions.

Providing a larger number of in-combinations in each group's in–out pattern raises a complication. During collection initialization, upon discovery of an out-combination that cannot be matched because the in-combination is already producing an out-combination needed to match a previous observed out-combination, another in-combination for that ID would be needed so that this new observed out-combination could be associated with this new in-combination. The idea is that this ID must have been reacting to this new, unobserved in-combination to generate this different out-combination.

This required, different in-combination could be provided by allowing each ID to generate two different out-combinations at each time point. These two different out-combinations would be viewed as different in-combinations at the next time point. A group ID would react to each of these in-combinations and post to the bulletin board two out-combinations. Both of these out-combinations would be examined and a match declared if the observed out-combination matched at least one of these model-generated out-combinations.

Allowing multiple out-combinations at each time step in the algorithm used to find the initial collection is the subject of future research.

11.3.2.3   Initialize step solution

Table 11.5 gives the results of the **Initialize** step.

Let entries in β_I be the modified values of those parameters operated on by the **Initialize** step – and the hypothesis value for all other parameters. The IntIDs

Table 11.5   **Initialize** step match statistics.

Group	$f^{(i)}$	$n_{uoc}^{(i)}$	$n_P^{(i)}/n_{iop}^{(i)}$
kenpres	0.130	1	0.875
kenepa	0.292	3	0.785
kenrr	0.333	3	0.958
kenpas	0.000	1	0.909
tanpres	0.200	2	1.000
tanepa	0.058	3	1.000
tanrr	0.428	3	0.800
tanpas	0.000	1	1.000
ugapres	0.076	2	0.666
ugaepa	0.027	3	0.578
ugarr	0.380	3	0.739
ugapas	0.000	1	0.909
ngo	0.042	2	0.760

model specified by the values in $\boldsymbol{\beta}_I$ matches 42 or 16.6% ($f = 0.166$) of the 252 action–target observations. The overall action match fraction is 21.8%, and the overall target match fraction is 59.5%.

The vector $\boldsymbol{\beta}_I$ is used as the starting point in the **Maximize** step, discussed next.

11.3.2.4   Maximize step solution

Hypothesis distribution agreement function Let

$$I^{(i)}(\boldsymbol{\beta}_I, \boldsymbol{\beta}_C) \equiv \frac{g_H^{(i)}(\boldsymbol{\beta}_I) - g_H^{(i)}(\boldsymbol{\beta}_C)}{|g_H^{(i)}(\boldsymbol{\beta}_I)|}. \tag{11.1}$$

For the group IDs, the value of $I^{(i)}(\boldsymbol{\beta}_I, \boldsymbol{\beta}_C)$ should be negative because the maximization procedure has found a set of parameter values, $\boldsymbol{\beta}_C$, for which the ID's distribution is closer in agreement to the distribution defined by $\boldsymbol{\beta}_H$ under the constraint that the match fraction generated by $\boldsymbol{\beta}_C$ is not smaller than the value generated by $\boldsymbol{\beta}_I$.

For the ecosystem ID, however, because $\boldsymbol{\beta}_I = \boldsymbol{\beta}_H$, $I^{(i)}(\boldsymbol{\beta}_I, \boldsymbol{\beta}_C) = I^{(i)}(\boldsymbol{\beta}_H, \boldsymbol{\beta}_C)$ and hence should be positive because the maximization procedure has found a set of parameter values, $\boldsymbol{\beta}_C$, that improves the agreement of the consistent distribution with the empirical distribution (the data) at the expense of agreement with the hypothesis distribution. Note that with the ecosystem ID, there is no match fraction constraint as there is with the group IDs.

The above holds only when there are a sufficient number of Monte Carlo realizations to allow the approximate computation of g_H measures to be close to their analytical values.

11.3.2.5   Group IDs

The truth tables of Section 11.2.3 require scenario goal node distributions to have particular patterns of values. To study how such required patterns force these nodes to have distributions different from their hypothesis values, for each group ID, parameters on only these nodes are allowed to be modified during the **Maximize** step of consistency analysis. Due to computational expense, only one such goal node per group ID has its parameters adjusted during this computation. In addition, only the first set of conditioning values was used to build the $g_H^{(Grp)}$ agreement measures.

For this analysis, hypothesis parameter values are set to represent the following theoretical statements (see also Appendix A):

1. The president believes that the only way to change a `Situation Maintain Political Power Goal` value of *unattained* to the value *attained* in the scenario is to win major approval from campaign donor audiences.

2. An EPA sees their chances of attaining their goal of expanding their budget when three conditions happen simultaneously: the state of this goal in the situation is close to being attained, the president in the scenario is more satisfied, and the scenario economic resources increase.

3. Rural residents see the chances of satisfying their families increase as the state of their family is better in both the situation and under a proposed action.

4. Pastoralists see the chance of being able to protect their livestock as increasing when their livestock are safe in the situation and safe under a proposed action.

5. Conservation-focused NGOs see their chance of maintaining good relations with host countries as increasing as this goal is increasingly attained in the situation, and it is perceived that presidents will be satisfied with a proposed action.

Table 11.6 gives the final consistency analysis agreement functions and summarizing measure of the cognitive theory's consistency with observations, n_{agree}/n_{dists} (see Chapter 4).

11.3.2.6 Ecosystem ID

For this ID, only parameters defining the cheetah birth rate, death rate, and abundance nodes are modified during the **Maximize** step. Also, only the Mara

Table 11.6 Consistency analysis agreement function values. See Chapter 4 for a discussion of the hypothesis agreement summary measure, n_{agree}/n_{dists}. The g_H values are given to allow comparisons to be made among group IDs.

Group	$g_H^{(i)}(\beta_C)$	$I^{(i)}(\beta_I, \beta_C)$	n_{agree}/n_{dists}	Scenario goal being estimated
kenpres	−0.2842	−0.229	1.0	Maintain political power
kenepa	−0.2581	−0.527	1.0	Expand budget
kenrr	−0.0952	−0.769	1.0	Support family
kenpas	−0.0509	−0.752	1.0	Protect livestock
tanpres	−0.0800	−0.140	1.0	Maintain political power
tanepa	−0.4296	−0.208	1.0	Expand budget
tanrr	−0.0440	−0.881	1.0	Support family
tanpas	−0.0246	−0.887	1.0	Protect livestock
ugapres	−0.0512	−0.091	1.0	Maintain political power
ugaepa	−0.0755	−0.624	1.0	Expand budget
ugarr	−0.0729	−0.800	1.0	Support family
ugapas	−0.0710	−0.651	1.0	Protect livestock
ngo	−0.1363	−0.242	1.0	Maintain relations

district of Tanzania is evaluated. This computation produces $g_S^{Eco}(\boldsymbol{\beta}_H) = -0.9781$, $g_S^{Eco}(\boldsymbol{\beta}_C) = -0.8687$, $g_H^{Eco}(\boldsymbol{\beta}_C) = -0.0633$, $n_{agree}/n_{dists} = 1.0$, and $I^{(i)}(\boldsymbol{\beta}_I, \boldsymbol{\beta}_C) = 1.92$. The sign of $I^{(Eco)}$ is as expected.

11.3.2.7 Actions history displays

Figure 11.1 portrays simulator output over the same time period as the observations. In this figure, a dash indicates an action that is not a reaction to a previous input action. An 'X' indicates where simulator output agrees with an observed action. Symbols on the ends of linked actions indicate the action's EMAT category: 'e' for economic, 'd' for diplomatic, 'm' for militaristic, and 's' for an action towards the ecosystem. See Chapter 2, Section 2.4, for further details on these abbreviations and this plot's interpretation. This actions history is displayed for just the year 2004 in Figure 2.4 of Chapter 2 in order to bring out the complexity of group interactions.

Symbols on the ecosystem history plot identify different combinations of country and ecosystem output node (cheetah abundance and herbivore abundance). The symbol 'x' denotes a cheetah abundance observation, and symbol 'o' denotes a herbivore abundance observation. As can be seen at the bottom of Figure 11.1, the fitted ecosystem ID produces cheetah and herbivore abundances that are similar to those in the ecosystem data set (see Chapter 10).

11.4   Conclusions and another collection initialization algorithm

The East African cheetah example demonstrates that an IntIDs simulator of a political–ecological process can be fitted to a political–ecological data set. This model estimation computation can be accomplished on a personal computer circa 2009 and hence is within the means of a modestly funded agency charged with making ecosystem management decisions.

As a statistical side note, consistency analysis is not derived from Bayesian statistical arguments so no effort needs to be expended by the analyst to justify a prior joint distribution of IntIDs model parameters. Most importantly, though, there is no need to confront the open problem of determining when the Markov chain Monte Carlo (MCMC) simulation has converged. For example, Smith (2007) states:

> Posterior summaries of model parameters are ultimately of interest in Bayesian analyses. These can be computed from MCMC chains, provided that the chains have converged to and provide representative samples from the joint posterior distribution. In all but the simplest of models, the joint posterior has a non-standard distributional form. Convergence to an unknown joint posterior cannot be proven, and hence diagnostic tests have been developed to identify MCMC output that has not converged to a stationary distribution. Since diagnostic tests do not provide proof of convergence, it is prudent to employ more than one when assessing the quality of samples from an MCMC algorithm.

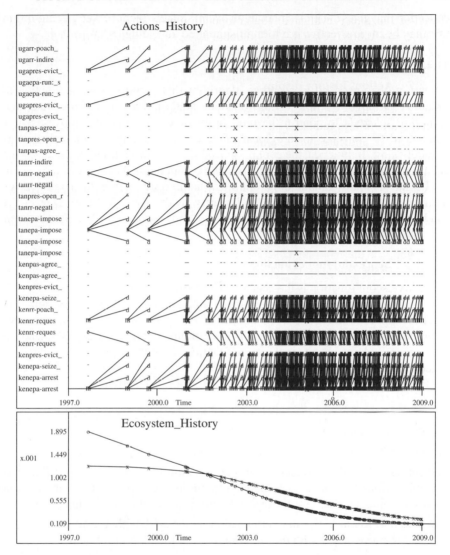

Figure 11.1   Simulator actions output under β_C values over the time period 1997 to 2009. In the actions history plot, an 'X' symbol denotes a model-generated action that matches an observed one. In the ecosystem time series plot, herbivore fraction detected and cheetah fraction detected are indicated by the 'o', and 'x' symbols, respectively.

11.4.1   Simulated annealing

It is well known that greedy search has a tendency to converge prematurely to a local maximum. Hence, the above algorithm may fail to find the in–out pattern collection that gives the global maximum overall match fraction. Another algorithm that may be more effective is a form of simulated annealing (SA).

11.4.1.1 Basic algorithm

SA is a probabilistic hill-descending algorithm that requires a *score* function to minimize, a *move* function to transform one state into a neighboring state, and a *cooling schedule* to control the probability of accepting uphill moves. SA is guaranteed to converge to the global minimum only if the cooling schedule is infinitely slow. In practice, the cooling schedule represents a trade-off between the chance of finding the global minimum and the computational expense of the search.

Unlike standard combinatorial optimization algorithms (see Taha 1975), SA's performance does not depend on the linearity of the score function. Hence, SA can be applied to problems that have a poorly understood and possibly nonlinear objective function.

11.4.1.2 Score and move functions

Here, SA's score function is the negative of the overall match fraction.

Design of the move function is critical. First, a move should be large enough so that the number of moves between any two states is not too big (limit the diameter of the state space). On the other hand, a move should not be so large that it leads to an excessively large change in the score function (keep the topography of the state space as smooth as possible).

Here, a move is defined to be the following four-step activity:

move-step 1: Randomly select a group ID.

move-step 2: Randomly select an in-combination.

move-step 3: Randomly select an out-combination from the list of possible out-combinations that includes a 'null' out-combination and excludes the current out-combination listed for the in-combination selected in move-step 2.

move-step 4: Replace the out-combination associated with the in-combination found in move-step 2 with the out-combination found in move-step 3.

Note that an in-combination can be associated with no more than one out-combination. Therefore, the above move function allows the number of in–out pairs to vary as the search progresses.

11.4.1.3 Finite time cooling schedule

See Aarts and Korst (1989, pp. 60–65) for the derivation of the cooling schedule and the proof that it will converge in a finite number of function evaluations. Because the number of evaluations is finite, the converged solution may not be the global minimum. The algorithm is as follows:

1. The current map of in-combinations to out-combinations (in–out pairs) is the initial solution with the addition that each in-combination not associated with any out-combination is mapped to the null out-combination.

2. The chain length, n_c, is equal to the size of the neighborhood, that is, the number of new states reachable in one move. Here, $n_c = \sum_{i=1}^{m} n_I^{(i)} n_O^{(i)}$. Note that the number of out-combinations to be chosen from is $(n_O^{(i)} + 1) - 1$.

3. Find an initial temperature, t, such that 91 to 99% of the moves are accepted.

4. At temperature t, compute a trial score function for n_c moves. Always accept a move if the score function is reduced, and accept an uphill move with probability $\exp[-(s_{\text{new}} - s_{\text{old}})/t]$. Calculate the average score, $\bar{s}$, and the score standard deviation, σ_s.

5. Consider the solution converged if $(t/\bar{s})[(\bar{s} - \bar{s}_{\text{old}})/(t - t_{\text{old}})] < \epsilon$ for a pre-specified value of ϵ.

6. If not converged, find a new temperature, $t_{\text{new}} = t/\left[1 + t \log(1 + \delta)/3\sigma_s\right]$. The distance parameter, δ, is user defined. Experience suggests 2.0 is a reasonable first guess for δ.

7. Go to 4.

The solution that yields the smallest value of the score function over all temperatures is the algorithm's output. This solution is not necessarily from the last chain.

11.5   Exercises

1. Suppose that five years have passed and additional cheetah abundance observations have been collected that exhibit a downward trend of 10% a year starting with the most recent abundance observation in the data set of Chapter 10. Perform a consistency analysis with this augmented political–ecological data set and explain any changes in fitted parameter values relative to those given above.

2. Rerun the analysis of Section 11.4 above, but do not allow any rural resident ID to have its parameters adjusted. Comment on how changes in the parameter values of other IDs compensated for this constraint.

3. Fit the stochastic climate model of Majda *et al.* (2009) to the Sea Surface Temperature data available from the Global Observing Systems Information Center (GOSIC 2010) using consistency analysis with c_H set in turn to 0.1, 0.5, and 0.9. Then compute a 95% prediction interval for average annual Sea Surface Temperature 50 years into the future.

4. Build a simple linear regression ID and use consistency analysis to fit it to the x, y data set: {(1, 1) (2, 3) (3, 2) (4, 4)}.

12

Assessing the simulator's reliability and improving its construct validity

12.1 Introduction

An EMT simulator is a stochastic model of a political–ecological system. As with any model that is to be used to help manage a real-world system (see Chapter 1), questions about the model's reliability and construct validity need to be answered. This chapter contains descriptions of procedures that are part of the politically realistic EMT that assess the reliability and construct validity of the EMT simulator. Herein, a *reliable simulator* is one that (a) does not have extreme sensitivity to any set of parameters, and (b) has predictive validity in that the simulator's predictions are superior to blind guessing. Therefore, to assess the reliability of a simulator under this two-dimensional definition of reliability, one needs to know the simulator's parameter sensitivities and predictive performance.

In addition to assessing the simulator's reliability, a procedure is needed for assessing the impact of different ID architectures and/or nodes on the simulator's construct validity (see Chapter 1). For the models described in this book, this amounts to seeing if the simulator's construct validity improves by changing all or parts of it in ways suggested by evolving theories of group decision making and wildlife population dynamics. Procedures derived from statistical hypothesis-testing theory can help with such assessments through their ability to compare the construct validity of two competing models.

Improving Natural Resource Management: Ecological and Political Models   Timothy C. Haas
© 2011 John Wiley & Sons, Ltd

Here, such two competing models have particular interpretations as detailed next. A central concept in statistical hypothesis testing is that of a *full model*. This is the model that the analyst believes contains all relevant predictor variables. To execute a hypothesis test, the analyst compares this full model to a reduced (nested) or alternative (nonnested) model. If the reduced or alternative model fits a data set about as well as the full model, the reduced or alternative model replaces the full model. See Weber and Skillings (2000, p. 127). The politically realistic EMT contains an MC hypothesis-testing procedure that can be used to perform such tests.

This chapter is structured as follows. First, a step-by-step procedure is given for assessing the reliability of a political–ecological system simulator. Second, to support one of these steps, an algorithm is given for performing a *sensitivity analysis* (see Saltelli 2000). Next, an algorithm is given for estimating the simulator's one-step-ahead prediction error rate. After that, instructions are given for using the EMT to conduct an MC hypothesis test of a proposed modification to one or more of the simulator's IDs. The last section contains a discussion of the effect on a hypothesis test that an unobserved covariate may have. Such an effect can occur when the data arises from an observational study rather than a designed experiment. A political–ecological data set in almost all cases arises from an observational study.

12.2   Steps for assessing simulator reliability

The following procedure can be used to assess a simulator's reliability and, if necessary, modify the simulator to improve its reliability. This procedure is based on the two-component definition of reliability given above.

Step 1: Fit the simulator with consistency analysis to a political–ecological data set (see Chapter 11).

Step 2: Perform a sensitivity analysis (see below). If the simulator is extremely sensitive to a set of parameters, reformulate the simulator and repeat step 1. Continue in this manner until the simulator is not excessively sensitive to any set of its parameters. A model is sensitive to a set of parameters if small perturbations to their values radically affect the model's outputs.

Step 3: Compute the simulator's prediction error rate. If it is worse than blind guessing, use the sensitivity analysis results to identify sets of parameters to which the simulator has low sensitivity. Temporarily leave these parameters out of the simulator and recompute the predictive error rate. If the error rate improves, permanently remove those parameters from the simulator and go back to step 1. Continue in this manner until the simulator's predictive performance is better than blind guessing.

The reason for this attempt at reducing model complexity is that in many model architectures, one cause of low predictive performance (a high prediction error rate) is model over parameterization.

Step 4: Use this reliable simulator to manage the ecosystem. Exhibit the simulator's reliability by posting the results from the final round of sensitivity analyses and predictive error rate computations on the EMT's website.

12.3   Sensitivity analysis

There are two main types of sensitivity analyses: probabilistic and deterministic (for a review see Marchand *et al.* 2008). Here, only a form of deterministic sensitivity analysis is studied due to the extreme computing expense of a probabilistic sensitivity analysis (see, e.g., Helton and Davis 2000, p. 111).

12.3.1   Rationale

A sensitivity analysis consists of computing measures of the amount of model output change due to perturbations of the model's parameters. Such small parameter changes could be due to *parameter misspecification*, that is, assigning a value to a parameter that is different than its true value. Failure to detect extreme sensitivities can seriously undermine the acceptance of simulator-based recommendations. Healy and Ascher (1995) note, for example, that critiques of the USDA Forest Service's reliance on FORPLAN output to arrive at forest management decisions were based inpart on studies that showed excessive sensitivity of FORPLAN's linear programming solutions to small changes in the values of parameters whose original values had been assigned with weak justification.

Another way of looking at this issue is to note that if a model's output is essentially determined by a few parameters but the subject matter theory on which the model is based makes no such claims, the model is not faithfully reflecting the subject matter theory from which the model was derived. In this case, the sensitivity analysis becomes a diagnostic for the model's construct validity. An example of a model with extreme sensitivity is the coupled system of differential equations used to forecast global economic status as described in the book *Limits to Growth* (Meadows *et al.* 1972). Vermeulen and De Jongh (1977) performed a sensitivity analysis of this model and found it to be unreliable due to its extreme sensitivity to a small number of parameters that could be assigned only poorly justified values. This sensitivity analysis effectively ended the influence that forecasts from this model had on the debate about the economic consequences of the world's population growth rate.

A third way to view the issue of parameter sensitivity is from the perspective of parameter value assignment. Parameters may have their values either assigned by experts or statistically estimated from a data set. Model sensitivity to expert-specified parameters needs to be assessed to determine how reliable such opinions need to be. Model sensitivity to statistically estimated parameters needs to be assessed so that data collection efforts can be focused on observing those variables represented by the model that have distributions defined by parameters to which the model has relatively high sensitivity.

12.3.1.1   One parameter versus sets of parameters

How model output is affected when only one parameter is perturbed while holding all other parameters at fixed values is referred to herein as *sensitivity to individual parameter misspecification*. This type of sensitivity is the typical focus of a sensitivity analysis: find parameter and/or intervals of values of input variables to which the model has high sensitivity. For a review of this type of sensitivity, see Saltelli (2000). A systematic method for finding the effect of individual parameters and/or input variables on a model's output is given in Morris (1991).

On the other hand, a model's sensitivity to small changes to a set of its parameters needs to be assessed to determine the model's overall robustness to the simultaneous misspecification of parameters in this *parameter set*. This type of sensitivity will be referred to herein as *sensitivity to parameter set misspecification*. In other words, this type of sensitivity is the size of the effect on output variables from joint perturbations of all or a subset of the model's parameters.

Note that this type of sensitivity cannot be assessed with any method that varies only one parameter at a time.

If a model does not radically change its output values under small changes to a subset of its parameters, the model is said to have *low sensitivity* to misspecification of these parameters.

12.3.2   Deterministic sensitivity analysis algorithm

An algorithm is developed here for a deterministic sensitivity analysis of an EMT simulator. Specifically, this algorithm computes measures of how sensitive the simulator is to misspecification of sets of parameters, that is, not simply misspecification of a single parameter. Let $out^{(i)}$, $i = 1, \ldots, k$, be functions of the simulator's output nodes. The algorithm is as follows:

Step 1: Execute a consistency analysis and update β_H to the resulting β_C.

Step 2: Define a scientifically significant change in simulator output to be a set of values on this set of output node functions $out_{DSA}^{(i)}$, $i = 1, \ldots, k$, where DSA stands for 'Deterministic Sensitivity Analysis.' Typically, a critic of the simulator would be asked to provide this set of values.

Step 3: Define an optimization objective function to be

$$f_{DSA}(\boldsymbol{\beta}) \equiv \sum_{i=1}^{k} |out_{DSA}^{(i)} - out^{(i)}(\boldsymbol{\beta})| + \sum_{j=1}^{g} \left(u_j(\boldsymbol{\beta}) - v_j(\boldsymbol{\beta}) \right) \qquad (12.1)$$

where $out^{(i)}(\boldsymbol{\beta})$ is the value of the ith output node function computed by the simulator under $\boldsymbol{\beta}$, and g is the number of group IDs in the simulator.

For the jth group ID, let

$$u_j(\boldsymbol{\beta}) \equiv \frac{\sum_O E_{\boldsymbol{\beta}}[\text{OGA}]}{\sum_T E_{\boldsymbol{\beta}_H}[\text{OGA}]} \qquad (12.2)$$

where OGA is that group ID's `Overall Goal Attainment` node, O is the set of out-combinations computed to be optimal by the group ID under $\boldsymbol{\beta}_H$, and T is the set of all out-combinations that the group ID is capable of producing. Also for this group ID, let

$$v_j(\boldsymbol{\beta}) \equiv \frac{\sum_{T-O} E_{\boldsymbol{\beta}}[\text{OGA}]}{\sum_T E_{\boldsymbol{\beta}_H}[\text{OGA}]} \qquad (12.3)$$

where $T - O$ denotes all out-combinations in T except those in O. The functions $u(\boldsymbol{\beta})$ and $v(\boldsymbol{\beta})$ are included in the f_{DSA} definition so that f_{DSA} exhibits a gradient in regions of the parameter space for which the group ID's out-combinations match those produced by the group ID under $\boldsymbol{\beta}_H$.

Starting at the point $\boldsymbol{\beta}_H$, search for a set of parameter values that minimizes $f_{DSA}(\boldsymbol{\beta})$. Call the solution to this optimization problem $\boldsymbol{\beta}_{DSA} = \arg\min_{\boldsymbol{\beta}} \{f_{DSA}(\boldsymbol{\beta})\}$.

Step 4: Find the parameter in $\boldsymbol{\beta}_{DSA}$ that is the least changed from its value in $\boldsymbol{\beta}_H$ – relative to its range of plausible values. Say that it turns out to be the lth parameter. Then $\beta^{(l)}$ is the most sensitive parameter and the difference, $|\beta_H^{(l)} - \beta_{DSA}^{(l)}|$, is the accuracy to which this parameter needs to be known. If $\beta_{DSA}^{(l)}$ is inside the 95% confidence interval for $\beta^{(l)}$, or $\beta_{DSA}^{(l)}$ is a plausible value for $\beta^{(l)}$, then this particular sensitivity analysis cannot be used as justification for dismissing the critic's concerns about the simulator's reliability.

The idea of this algorithm is to search for a new set of parameter values that is as close to $\boldsymbol{\beta}_H$ as possible but causes the simulator's outputs to change by an amount that is scientifically significant.

This algorithm implements two perspectives on how a sensitivity analysis should proceed: (a) a purposeful search should be conducted for a set of parameter values that causes the simulator to deviate significantly from the behavior it exhibits under parameter values estimated from data; and (b) a sensitivity analysis is most relevant when it is carried out in regions of the parameter space that are supported by observations. This second perspective is why, in the EMT, a sensitivity analysis is conducted only after the simulator has been fitted to data through consistency analysis.

In effect, this sensitivity analysis algorithm is 'tweaking' the simulator until its behavior changes in a scientifically significant way. Hence, this approach to sensitivity analysis directly addresses criticisms that the simulator may have been 'tweaked' to give output that the analyst has a priori decided should be the case. This approach also directly handles the difficult question of sensitivity to misspecification of several parameters simultaneously rather than to one parameter at a time.

There may be points in the parameter space even closer to β_H that would result in the model producing the values $out_{DSA}^{(1)}, \ldots, out_{DSA}^{(k)}$ but have been skipped over during the course of the optimization run. In this case, the model's sensitivity is even worse than computed. Hence, this algorithm yields an optimistic assessment of a model's sensitivity to parameter set misspecification.

12.3.2.1 Conceptual example

One of the controversies surrounding the global warming issue is the sensitivity of Global Climate Models (GCMs) to parameter misspecification (see, e.g., Murphy *et al.* 2004). For this case, the above deterministic sensitivity analysis would proceed as follows. First, a particular GCM would be selected, say the Integrated Global System Model maintained by the Center for Climate Change at the Massachusetts Institute of Technology (MIT) (see Sokolov *et al.* 2009).

Next, a scientifically significant change in the GCM's output would be developed by noting that critics of global warming dispute the claim that the earth's average annual temperature will rise 50 years hence. Instead, these critics view a zero degree increase in annual average temperature as an equally plausible outcome. Let $out_{DSA}^{(i)}$ be the vector of the change in annual mean sea surface temperature between the present day and 50 years into the future at the ith location in the North Atlantic Ocean. Say that $k = 100$ locations across this ocean are selected. Then the hypothesis of no global warming can be represented by setting $out_{DSA}^{(i)} = 0$, $i = 1, \ldots, k$.

Finally, the above sensitivity analysis algorithm would be run. If, during execution of this algorithm, the model's parameters are adjusted to a set of scientifically plausible parameter values for which $f_{DSA}(\beta) = 0$, then this climate model predicts global warming under one set of plausible parameter values – and predicts no global warming under another set of plausible parameter values. In other words, the GCM exhibits high parameter sensitivity.

12.3.3   Sensitivity analysis example

The above sensitivity algorithm is run on the East African cheetah EMT simulator over the time interval of January 1, 1997 to December 31, 2009.

12.3.3.1 Definitions of sensitivity measures

In this example, let n_{OC} be the total number of observed out-combinations in the political–ecological data set of Chapter 11 over this time interval. Let n_{eco} be the number of observations on an ecosystem ID node of interest. The simulator is run under β_H and computes an out-combination for each group and time point in this data set. This set of out-combinations is stored. Let $n_{matched}^{(OC)}(\boldsymbol{\beta})$ be the number of these out-combinations that are matched by the simulator under $\boldsymbol{\beta}$. Let $E_{\boldsymbol{\beta}_H}[N_{t_i}]$ be the expected cheetah abundance at an observed time t_i computed by the simulator under $\boldsymbol{\beta}_H$. Define $E_{\boldsymbol{\beta}}[N_{t_i}]$ similarly.

Using these components, define one output measure for the group IDs, and one output measure for the ecosystem ID, as follows:

$$out^{(1)}(\boldsymbol{\beta}) = (n_{OC} - n_{matched}(\boldsymbol{\beta}))/(n_{OC} + 1) \qquad (12.4)$$

and

$$out^{(2)}(\boldsymbol{\beta}) = \sum_{i=1}^{n_{eco}} (E_{\boldsymbol{\beta}_H}[N_{t_i}] - E_{\boldsymbol{\beta}}[N_{t_i}])/(E_{\boldsymbol{\beta}_H}[N_{t_i}] + 1). \qquad (12.5)$$

The first component of **out** is the number of times an out-combination generated by the simulator under $\boldsymbol{\beta}$ does not match the out-combination generated by the model fitted to data with consistency analysis (as represented by $\boldsymbol{\beta}_H$) – relative to the total number of out-combinations generated by this statistically estimated simulator. The second component of **out** is the sum of relative changes in the expected value of the ecosystem ID's N_t node across all observation times.

If small changes to $\boldsymbol{\beta}_H$ can cause scientifically significant values of either of these components, then the model has excessive sensitivity to some set of its parameters. Note that by these definitions, the simulator's sensitivity is being assessed only at time points for which a group or the ecosystem has been observed. In particular, say that $out_{DSA}^{(i)} = .1$, $i = 1, 2$; that is, an out-combination match rate that is less than 90% indicates the simulator under $\boldsymbol{\beta}$ is scientifically significantly different from the simulator under $\boldsymbol{\beta}_H$. A similar statement holds for N_t.

12.3.3.2   Optimization runs

For the first run, only parameters defining the scenario `Relative Military Resource Change` node within the Kenya president ID were allowed to vary. This run of the algorithm converged and produced an $out^{(1)}$ value of .1194. The smallest change in a parameter value was 4 – from *large decrease* (the value 1) to *large increase* (the value 5) and viceversa. This change resulted in the number of matched out-combinations falling from 200 to 177. This result indicates the model is quite robust to misspecification of the parameters that define the `Relative Military Resource Change` node.

For the second run, only parameters defining the cheetah abundance node, N_t, were allowed to vary (see Chapter 8 for definitions of these parameters). For this second run, the algorithm also converged and produced an $out^{(2)}$ value of .1102. The parameters with the smallest percentage changes were N_0 (which changed from 200 to 360) and P (which changed from .1 to .18). As these are both 80% changes, both of these parameters are also robust to misspecification.

12.4   One-step-ahead prediction error rates

Haas (2008b) gives an early version of the error rate algorithm described below.

To establish the second aspect of reliability as laid out in the Introduction, the simulator's prediction error rate needs to be significantly lower than the prediction error rate of blindly guessing what actions will be taken by groups and what effect such actions will have on the ecosystem.

Group IDs in the simulator produce nominally valued output in the form of out-combinations. The ecosystem ID produces continuously valued output in the form of expected wildlife abundance values. Two different measures of prediction error rates, then, are needed.

For action–target output, the prediction error rate will be called herein the One-Step-Ahead Predicted Actions Error Rate (OSAPAER), and for the continuously valued ecosystem output, the prediction error rate will be called the One-Step-Ahead Root Mean Squared Prediction Error (OSARMSPE).

12.4.1   One-step-ahead predicted actions error rate

One way to define the OSAPAER is as follows. Consider a large but finite number of sequential time points, $t_1, \ldots, t_T$. At each of these time points, one or more of the simulator's group IDs posts an out-combination. Then, let

$$\text{OSAPAER} \equiv \frac{1}{T} \sum_{i=1}^{T-1} 1 - \frac{n_{matched}^{(i+1)}}{n_{observed}^{(i+1)}} \qquad (12.6)$$

where $n_{matched}^{(i+1)}$ is the number of out-combinations generated by the simulator at time point t_{i+1} that match observed out-combinations at that time point, and $n_{observed}^{(i+1)}$ is the number of these observed out-combinations.

One way to estimate this error rate can be described in general terms as follows. First, re fit the simulator at each time point in the political–ecological data set using all data up to but not including that time point. Then, at each of these time points, use this refitted simulator to compute a prediction of each group's out-combination and a prediction of the ecosystem's state. Finally, by comparing these one-step-ahead predictions to the observed values, produce an estimate of the one-step-ahead prediction error rate. The idea motivating this is that, in actual use, an ecosystem manager would always refit the simulator as soon as new actions and/or new ecosystem variable values were observed before using the simulator to predict future group actions and/or future ecosystem variable values.

More specifically, to compute this estimate, one proceeds by starting back n_{pred} time points from the latest time point in the data set (T_D), and then performing the following two computations at each of the time points $T_D - n_{pred} + i$, $i = 0, \ldots, n_{pred} - 1$. First, the simulator is refitted using all observed out-combinations up through time $T_D - n_{pred} + i$. Then, this refitted simulator is run from the first time point in the data set up through time point $T_D - n_{pred} + i + 1$. After all these computations are completed, one may estimate the OSAPAER with

$$\widehat{\text{OSAPAER}} = \frac{1}{n_{pred}} \sum_{i=T_D-n_{pred}}^{T_D-1} 1 - \frac{n_{matched}^{(i+1)}}{n_{observed}^{(i+1)}} \tag{12.7}$$

where $n_{matched}^{(i+1)}$ is the number of out-combinations generated by the simulator at time point t_{i+1} that match observed out-combinations at that time point, and $n_{observed}^{(i+1)}$ is the number of these observed out-combinations.

To reduce the expense of computing $\widehat{\text{OSAPAER}}$, simulator refitting can be performed only at every lth time point. For example, if $l = 3$, the simulator would be refitted only at time points $T - n_{pred}$, $T - n_{pred} + 3$, $T - n_{pred} + 6$, $T - n_{pred} + 9$, ..., $T - n_{pred} + 3m$ where $m = floor(n_{pred}/3)$. As another aid to reducing computational expense, the **Maximize** step of consistency analysis need not be run on group IDs because the number of matches between observed and model-generated actions only depends on the simulator's collection of in–out patterns, which, in turn, is essentially set at the end of the consistency analysis **Initialize** step.

Say that a group ID has K decision options (out-combinations). In the worst case, one of these options has a high probability of being chosen at each time point. Blind guessing, that is, assuming all options are equally likely, would predict this option with probability $1/K$ at each time point resulting in an error rate of about $1 - 1/K$. An ecosystem manager would prefer the simulator's predictions over blind guessing whenever $\widehat{\text{OSAPAER}} < 1 - 1/K$.

12.4.2 One-step-ahead root mean squared predicted error rate of ecosystem values

When the quantity to be predicted is continuously valued, a commonly used measure of predictive accuracy is the root mean squared prediction error. Using the notation of the previous section, let

$$\text{OSARMSPE}_i \equiv \left[\frac{1}{T} \sum_{j=1}^{T-1} \left(Y_{i,j+1}^{(obs)} - Y_{i,j+1}^{(pred)} \right)^2 \right]^{1/2} \tag{12.8}$$

where $Y_{i,j+1}^{(obs)}$ is the observed value of the ith ecosystem output variable at time point $j + 1$, and $Y_{i,j+1}^{(pred)}$ is the ecosystem ID's prediction of this variable at time point $j + 1$ after being fitted to data up through time point j.

For ecosystem ID nodes, the OSARMSPE can be estimated with

$$\widehat{\text{OSARMSPE}}_i = \left[\frac{1}{n_{pred}} \sum_{j=T_D-n_{pred}}^{T_D-1} \left(Y_{i,j+1}^{(obs)} - Y_{i,j+1}^{(pred)} \right)^2 \right]^{1/2} \tag{12.9}$$

where T_D and n_{pred} are as defined in the previous section.

12.4.3   An example of computing prediction error rates

The $\widehat{\text{OSAPAER}}$ is computed for the East African cheetah EMT simulator. For this example, T_D is set to 2009.0, and actions for the years 2007 and 2008 are predicted ($n_{pred} = 37$). The simulator is refitted every 20 time points, which, for this example, results in running the **Initialize** step of consistency analysis at time point 2006.92 and then again at 2007.58, each time using data only up to that time point. The resulting parameter values are used to compute predictions of all 36 actions observed in the year 2006. These predictions give an $\widehat{\text{OSAPAER}}$ value of .5, or six actions being predicted correctly.

The average number of out-combinations across all group IDs is about 18 – giving a blind guessing error rate of .94. Hence, this conservative estimate of the simulator's $\widehat{\text{OSAPAER}}$ value indicates that the simulator is significantly more accurate than blind guessing.

The ecosystem ID is run over the Tanzanian regions of Mara, Rukwa, and Shinyanga. Table 12.1 contains the three pairs of observed and predicted values that lead to an OSARMSPE value on the cheetah nonsurvey sightings node of 17.32. The ratio of the OSARMSPE to $\bar{y}_i^{(obs)}$ is .7217 where $\bar{y}_i^{(obs)}$ is the mean of the observed values on the ith variable.

Table 12.1 Observed and one-step-ahead predictions on the cheetah nonsurvey sightings node.

Region	Year	Observed	Predicted
Mara	2008	30	0
Rukwa	2008	9	0
Shinyanga	2008	33	0

12.5   MC hypothesis tests

In what follows, the East African cheetah EMT simulator of Chapters 7 and 8 can be considered the full model in the language of hypothesis testing (see Section 12.1, above).

To build models of political–ecological processes, this author recommends the *Probabilistic Reduction Approach* (PRA) proposed by Spanos (1995) and reviewed by Kaplan (2009, pp. 212–220). This model-building procedure consists of three steps:

1. Articulate in sufficient detail a theoretical explanation of the behaviors of the random variables to be observed. In structural equation models (SEMs), for example, this entails specifying relationships between the X variables (inputs) and the Y variables (outputs).

2. Fit a *statistical model*, that is, a full model in linear models theory, or a saturated model in loglinear models theory. A statistical model is a joint probability distribution of the observable random variables that is not significantly different from the stochastic process that is being observed.

   To establish this property of nonsignificant difference between the actual process and the statistical model, the observable random variables need to be sampled and then statistical tests performed to assess the fit of the model's joint distribution to the data.

3. Place restrictions on the parameters of this statistical model to represent the theoretical model delineated in step 1. Typically, this involves specifying some parameters to be zero, others to be positive, and still others to be negative. Then, perform a hypothesis test to see if these restrictions cause this reduced model to be significantly different from the statistical model.

When asked what could be done with observational study data to help establish causality, the statistician Sir Ronald Fisher responded with 'Make your theories elaborate' (as quoted by Cochran (1965) and reported in Rosenbaum (1995)). From the viewpoint of a statistician, the simulator of Chapters 7 and 8 is an elaborate model.

A hypothesis-testing procedure is needed, then, that can be used to explore theoretical questions about group decision making and ecosystem dynamics. One such procedure, described in the next section, is based on the model's *out-of-sample* error rate. The out-of-sample error rate is the model's ability to correctly predict observations that are not contained in the sample that was used to fit the model. This out-of-sample error rate is different from a model's one-step-ahead prediction error rate. One way to estimate the out-of-sample error rate is with a *delete-d jackknife* estimator, described next.

12.5.1   Out-of-sample error rate

The following is based on Hand (1997, pp. 120–121). Say that a size n sample, s_n, is drawn from the joint distribution of the K-valued response variable, $Y(\mathbf{x})$, and the predictor variables (covariates), $\mathbf{X}$. This sample contains n_k observations on the kth out-combination with $n = \sum_{k=1}^{K} n_k$. Let

$$\mathbf{s}_n = (\mathbf{x}_{1,1}, \dots, \mathbf{x}_{1,n_1}, \dots, \mathbf{x}_{K,1}, \dots, \mathbf{x}_{K,n_K})' \tag{12.10}$$

where $\mathbf{x}_{k,i}$ is the ith design point at which the value k of the response variable is observed. Let $R_k^{(\mathbf{s}_n)}$ be the region of the predictor space for which the classifier will predict value k after having been constructed with $\mathbf{s}_n$.

The actual error rate (AER) is

$$\sum_{k=1}^{K} \int_{R_k^{(\mathbf{s}_n)}} [1 - \hat{P}(Y = k|\mathbf{x})] f(\mathbf{x}) d\mathbf{x} \tag{12.11}$$

where $P(Y = k|\mathbf{x})$ is the response variable's probability mass function at $\mathbf{x}$, and $\hat{P}$ denotes this distribution has been estimated from $\mathbf{s}_n$. The AER is conditional on $\mathbf{s}_n$ and can be interpreted as the in-sample error rate computed by applying the classifier built from $\mathbf{s}_n$ to an infinitely large sample (not $\mathbf{s}_n$) from the joint distribution between the random variable, $Y(\mathbf{x})$, and the random vector, $\mathbf{X}$.

The expected error rate (EER) removes this conditioning on $\mathbf{s}_n$ by taking the expected value of the AER over the distribution of $\mathbf{S}_n$:

$$\text{EER} = \int \text{AER} f_{\mathbf{S}_n}(\mathbf{s}_n) d\mathbf{s}_n. \tag{12.12}$$

Note that the EER is a deterministic function of the parameters defining the response variable's distribution and the joint distribution of $\mathbf{X}$. Hence, the EER can be thought of as a parameter itself.

12.5.1.1   Estimating the AER of simulator IDs

Let $\widehat{\text{AER}}$ be an estimate of AER. The *delete-d jackknife* (Shao and Tu 1995, Politis and Romano 1994) can be applied to the time series data of Chapters 9 and 10, that is, political–ecological data, to define one form of $\widehat{\text{AER}}$ as follows:

1. Randomly select a set of d observations and temporarily hold them out of the political–ecological data set.

2. Statistically fit the simulator's parameters with the remaining $r \equiv n - d$ observations.

3. At the time, region, group or ecosystem point of each of the d held-out observations, predict the held-out observation's value. For a group ID, estimate the AER to be the number of mismatches over these d predictions. For ecosystem ID output variables, compute the RMSPE of these d predictions:

$$\widehat{\text{RMSPE}}_i = \left[\frac{1}{d} \sum_{j=1}^{d} \left(Y_{i,j}^{(obs)} - Y_{i,j}^{(pred)} \right)^2 \right]^{1/2}. \tag{12.13}$$

Note that these two error rates are computed with a simulator that has been fitted once to r observations, some of which may carry a time stamp that is earlier than that of a held-out observation, and some of which may carry a time stamp that is later than a held-out observation. Hence, these two error rates are not the same as the OSAPAER and OSARMSPE statistics of Section 12.4, above.

Politis and Romano (1994) describe such a delete-d resampling procedure to form subsamples from a times series data set. Chapters 9 and 10 contain descriptions of the sampling assumptions of the group IDs and the ecosystem ID, respectively. Basically, the actions history data, although not viewed as being generated by a random process, has sampling variability due to its collection protocol – and the ecosystem data has sampling variability because ecosystem variables are modeled as a multivariate spatio-temporal stochastic process. Therefore, the entire political–ecological data set is viewed as a sample from a multivariate stochastic process for which the discrete variables are deterministic (actions and targets), and the continuous variables are stochastic (ecosystem variables). Hence, the assumptions of Politis and Romano (1994) apply to the process that generates political–ecological data.

12.5.2   MC hypothesis test theory

Typically, a hypothesis test is performed on the hypothesis that the effect on the response variable of a set of covariates is exactly zero. Such a test can be constructed with a computer-intensive approach called an *MC hypothesis test*. One way to construct such a test is to use *resampling statistics*. See Hall and Titterington (1989) for a description and asymptotic properties of MC hypothesis tests, and see Shao and Tu (1995) for a similar discussion of resampling statistics.

Let the null hypothesis, H_0, be that there is no difference in the out-of-sample error rate between a reduced simulator and the full simulator. Let $\delta \equiv \text{AER}_R - \text{AER}_F$ where AER_R and AER_F are the true error rates of the reduced simulator and the full simulator, respectively. Define the test statistic to be $T \equiv \hat{\delta} - \delta$. Note that under H_0, $T = \hat{\delta}$.

The AER is used as the basis for the test statistic because, in this author's opinion, for the purposes of both understanding a political–ecological process and predicting its behavior, the most important property of a simulator of such a process is its out-of-sample prediction accuracy. That is, if performance on this criterion is not acceptable, the simulator should not be used to make ecosystem management decisions.

Consider a group ID that has a repertoire of K out-combinations. If the observed frequencies across unique out-combinations are very different from one another, a possible drawback to using the $\widehat{\text{AER}}$ as the test statistic is that the test may be overly sensitive to error rate changes in those out-combinations that are the most frequent in the sample. This may result in declaring as insignificant covariates that in actuality significantly affect the error rate of the rarer out-combinations. Therefore, a second test statistic is defined based on the computation of the actual error rate sum (AERS) defined to be AERS $\equiv \sum_{i=1}^{K} \text{AER}_i$ where AER_i is the actual error rate for the ith out-combination. This summative error rate is used to define $\delta_S = \text{AERS}_R - \text{AERS}_F$, and finally $T_S = \hat{\delta}_S - \delta_S$. The T_S statistic is uniformly sensitive to changes across individual out-combination error rates no matter what the observed out-combination frequencies are and hence can be said to be less sample dependent than T.

12.5.2.1   MC hypothesis test procedure

The MC hypothesis test procedure with T as its test statistic is as follows:

Step 1: Compute $\widehat{\text{AER}}_R$ and $\widehat{\text{AER}}_F$ from the complete data set (hereafter called the *complete* sample). Compute $T = \hat{\delta}$, the observed value of the test statistic assuming H_0 is true.

Step 2: Sample without replacement r $(< n)$ observations from the complete sample. Call this subsample a *jackknife sample*.

Step 3: Compute reduced and full simulator error rate estimates using this jackknife sample. Denote these two error rate estimates AER_R^* and AER_F^*, respectively. Compute and store $T^* = \hat{\delta}^* - \hat{\delta}$, the jackknife sample's test statistic value. Note that the true (but unknown) error rates have been replaced with those estimated from the complete sample. Doing so improves the MC hypothesis test's power (Hall and Titterington 1989).

Step 4: Repeat steps 2 and 3 n_{MC} times (always with a newly drawn jackknife sample). The idea here is to form an empirical distribution of T.

Step 5: Compute the p value of the test to be the fraction of T^* values greater than T.

Notes:

1. When $r < n - 1$, the histogram of the n_{MC} T^* values forms a delete-d jack-knife statistic (Shao and Tu 1995, p. 197). For the hypothesis test to be consistent, both d and n_{MC} need to be large (Shao and Tu 1995, pp. 52 and 199). More specifically, Politis and Romano (1994) show that under weak conditions, as long as $r \to \infty$ while $r/n \to 0$, a statistic computed with delete-d jackknife samples is consistent. In particular, this result means that the confidence level of a delete-d confidence interval is not systematically biased for the true coverage probability. One way to guarantee these conditions is to have $r = n^\tau$ where $\tau \in (0, 1)$.

2. Efron (1983) finds the cross-validation (delete-1) $\widehat{\text{AER}}$ to be nearly unbiased for the AER. It is not clear how well Efron's results generalize to other values of d. The effect of such potential bias, however, is minimized by defining the test statistic to be a function of only *differences* in estimated error rates – as is done here with the definitions of T and T_S.

12.5.3 Hypothesis-testing example setup

To illustrate how hypothesis testing is used to improve a group ID's construct validity, the Kenya president ID could be reduced by removing the `Audience Satisfaction` node. The out-of-sample error rate of this reduced ID would be computed over $n_{MC} = 1000$ delete-d jackknife samples taken from the actions history data set in Chapter 9.

The test would be performed by executing the following steps:

1. Set $r = \lfloor \sqrt{n} \rfloor = \lfloor \sqrt{252} \rfloor = 15$ and $d = n - r = 252 - 15 = 237$.

2. Conduct the test using **id**'s `mc_hypothesis_test` relation of the `report` word's `sensitivity_analysis` qualifier.

12.6 Sensitivity to hidden bias analysis

12.6.1 Overview

A hypothesis test is ideally performed on data arising from a designed experiment. But a political–ecological data set is almost always a sample from an observational study. With such data, a significant test can be caused by an unobserved predictor variable as explained below. Analysis, therefore, is needed of the sensitivity of the test result to this unobserved variable. A review of such analyses is discussed next followed by a description of how such an analysis can be performed on a group ID within an EMT simulator.

12.6.2 Review of hidden bias sensitivity

Only when the analyst can dictate the chance that a subject is assigned to a treatment (usually by forcing these chances to be equal for all subjects) can the effect of unobserved predictor variables or *hidden* covariates be ignored. This is the case for a designed experiment. If, however, treatment assignment chances are not known, the resultant data set is said to have come from an *observational study*. In this case, the effect of possible hidden covariates on output variables can only be guessed at. Such guessing is referred to as performing an analysis of *sensitivity to hidden bias* (see Rosenbaum 2002, Chapter 4).

Say that there are M possible subjects available for the study, that is, the *population* size is M. Index the value of each subject's covariate vector with $\mathbf{x}_{[j]}$, $j = 1, \ldots, M$, and the associated treatment assignment variable with $Z_{[j]}$. Subject $[j]$ receives the treatment if $Z_{[j]} = 1$. Let $\pi_{[j]} = P(Z_{[j]} = 1)$, and for the entire population, $P(Z_{[1]} = z_1, \ldots, Z_{[M]} = z_M) = \prod_{j=1}^{M} \pi_{[j]}^{z_j}(1 - \pi_{[j]})^{1-z_j}$. Note that if $z = 1$, $\pi^z(1 - \pi)^{1-z} = \pi$, and if $z = 0$, $\pi^z(1 - \pi)^{1-z} = 1 - \pi$.

Say that N of the available M subjects are obtained and stratified on $\mathbf{x}$ into S strata such that $\mathbf{x}_{si} = \mathbf{x}_{sj}$ for all i and j in stratum s. Then, there are m_s subjects in stratum s that received the treatment.

Say that a hidden covariate, u, significantly affects the treatment assignment probabilities of subjects; that is, if subject $[j]$ and subject $[k]$ have the same values on the observed covariates, $\mathbf{x}$, but different values on u, then $\pi(\mathbf{x}_{[j]}, u_{[j]}) \neq \pi(\mathbf{x}_{[k]}, u_{[k]})$.

In this case, the ratio of the odds of subject j being assigned the treatment to the odds of subject k being assigned the treatment will be contained in the interval (Γ^{-1}, Γ) where $\Gamma > 1$. For example, for $\Gamma = 2$, this interval is $(1/2, \ 2)$. Γ is interpreted as the worst case scenario of this odds ratio.

Consider a binary logistic regression model for the treatment assignment choice:

$$\ln\left[\frac{\pi(\mathbf{x}_{[j]}, \ \mathbf{u}_{[j]})}{1 - \pi(\mathbf{x}_{[j]}, \ \mathbf{u}_{[j]})}\right] = \ln\left[\frac{\pi_{[j]}}{1 - \pi_{[j]}}\right] = \kappa(\mathbf{x}_{[j]}) + \gamma u_{[j]} \qquad (12.14)$$

where $0 \leq u_{[j]} \leq 1$, $\kappa(.)$ is an unknown function, and γ is an unknown parameter.

The above odds ratio can then be written as $\exp(\gamma(u_{[j]} - u_{[k]}))$. The extreme values of the odds ratio usually occur when u is either 0 or 1.

The researcher does not know a subject's u value. Hence, a sensitivity analysis consists of taking extreme values of u and then computing p values of the test statistic under different values of Γ. Since Γ is an odds ratio boundary, the researcher can state how much change in treatment assignment probability that an unobserved covariate would have to cause in order to change the p value from a significant value (say $< .05$) to an insignificant value (say $> .05$).

Therefore, sensitivity to hidden bias is expressed as how big Γ must be before significant p values become insignificant. The relationship between a p value and Γ could be plotted.

12.6.3   Matched pairs on x

Forming matched pairs is equivalent to stratifying on $\mathbf{x}$. One has a matched pair when each member in a pair has the same chance of being given the treatment: $\pi(\mathbf{x}_{si})$ for the ith member of the pair (stratum) s ($i = 1$ or 2). Here, $\mathbf{x}$ is the vector of observed covariates that are being used to define the strata.

A *propensity score* is simply an estimate of $\pi(\mathbf{x}_{si})$. Logistic regression can be used to estimate this probability as a function of $\mathbf{x}$. Then, pairs (strata) can be found as follows:

(a) Partition the unit interval into the subintervals $(0, \ p_1)$, $(p_1, \ p_2)$, $\ldots$, $(p_{S-1}, \ p_S)$.

(b) Assign a subject to be the first member of stratum s and another subject to be its second member if

   (i) exactly one of these subjects was given the treatment,

(ii) $\pi(\mathbf{x}_{s1}) \in (p_{s-1}, \ p_s)$, and

(iii) $\pi(\mathbf{x}_{s2}) \in (p_{s-1}, \ p_s)$.

All of this effort to find matched pairs is performed so that the conditional treatment–assignment probabilities are known (as in a randomized experiment) – not necessarily to make the members of a pair (stratum) to be in some sense similar to each other (Rosenbaum 2002, p. 83).

In a multiple regression analysis of an observational study, because the variables in $\mathbf{x}$ are continuously valued, $\mathbf{x}_{s1}$ may not exactly equal $\mathbf{x}_{s2}$. In this case, the use of logistic regression to achieve at least similar treatment–assignment probabilities within a stratum is needed – and is exactly what was done in Rosenbaum (1986), discussed next.

12.6.4   Hidden bias in multiple regression analyses

Rosenbaum (1986) studies the effect that dropping out of high school in the USA may have on a person's performance on intelligence tests. Clearly, a randomized experiment is not feasible to perform in this case. Rosenbaum (1986) therefore shows how to use an observational study data set to assess the effect of a qualitative predictor variable on a quantitative dependent variable when other quantitative or qualitative predictor variables are also measured. This situation is not dealt with in Rosenbaum's (2002) book. There are two typos in equation (8) of Rosenbaum (1986): the first conditional expected value on lines 1 and 2 should have $U = u_1$.

Rosenbaum's (1986) equation (7) implies that the data is being used to estimate the regression model:

$$R_0 = \alpha_s + \mathbf{x}'\boldsymbol{\beta} + \epsilon, \ \epsilon \sim N(0, \ \sigma^2), \tag{12.15}$$

for those who stay in high school ('stayers'), and

$$R_1 = \alpha_s + \Delta + \mathbf{x}'\boldsymbol{\beta} + \epsilon \tag{12.16}$$

for the 'dropouts' where α_s is the regression model constant for school s. The response is a person's score on an intelligence test.

Say that a hidden covariate, u, has some effect on R. In this case, the above two models are modified to be

$$R_0 = \alpha_s^* + \mathbf{x}'\boldsymbol{\beta}^* + u\phi + \epsilon \tag{12.17}$$

for the 'stayers,' and

$$R_1 = \alpha_s^* + \Delta^* + \mathbf{x}'\boldsymbol{\beta}^* + u\phi + \epsilon \tag{12.18}$$

for the 'dropouts.'

Consider the ith matched pair, $i = 1, \ldots, n$. Let $\mathbf{R}_1 = (R_{11}, \ldots, R_{1n})'$ be the vector of the 'dropout' member responses over these matched pairs. Define $\mathbf{R}_0$

similarly for the 'stayers.' For the ith matched pair, let $D_i = R_{1i} - R_{0i}$, $\mathbf{v}_i = \mathbf{x}_{1i} - \mathbf{x}_{0i}$, and $w_i = u_{1i} - u_{0i}$. Then

$$
\begin{aligned}
D_i &= \Delta^* + (\mathbf{x}_{1i} - \mathbf{x}_{0i})'\boldsymbol{\beta}^* + (u_{1i} - u_{0i})\phi + \epsilon \\
&= \Delta^* + \mathbf{v}_i'\boldsymbol{\beta}^* + w_i\phi + \epsilon.
\end{aligned}
\tag{12.19}
$$

Let $\mathbf{D} = \mathbf{R}_1 - \mathbf{R}_0$, $\mathbf{w} = (w_1, \ldots, w_n)'$,

$$
X = \begin{bmatrix} 1 & \mathbf{v}_1' \\ \cdot & \cdot \\ \cdot & \cdot \\ \cdot & \cdot \\ 1 & \mathbf{v}_n' \end{bmatrix},
\tag{12.20}
$$

and $p = \dim\{(\Delta\ \boldsymbol{\beta}')'\}$. Then

$$
\mathbf{D} = X \begin{bmatrix} \Delta \\ \boldsymbol{\beta} \end{bmatrix} + \epsilon
\tag{12.21}
$$

when there is no hidden covariate, and

$$
\mathbf{D} = (X\ \mathbf{w}) \begin{bmatrix} \Delta^* \\ \boldsymbol{\beta} \\ \phi \end{bmatrix} + \epsilon
\tag{12.22}
$$

when there is.

If data on u was available, this model could be estimated to yield parameter estimates denoted by $\hat{\alpha}_s^*$, $\hat{\Delta}^*$, $\hat{\boldsymbol{\beta}}^*$, and $\hat{\phi}$.

Say that a new predictor variable, w, is added to a multiple regression model. Further, assume that data on this new variable is acquired after the original model has been estimated. Let 'update' denote parameter estimates that are updated using data on this new predictor variable. For this situation, Seber's (1977, p. 66) Theorem 3.7 gives:

(a) $\hat{\phi} = (\mathbf{w}'F\mathbf{w})^{-1}\mathbf{w}'F\mathbf{D}$ where $F = \mathbf{1} - X(X'X)^{-1}X'$, and

(b)

$$
\begin{bmatrix} \hat{\Delta} \\ \hat{\boldsymbol{\beta}} \end{bmatrix}_{\text{update}} = \begin{bmatrix} \hat{\Delta} \\ \hat{\boldsymbol{\beta}} \end{bmatrix} - \hat{\boldsymbol{\eta}}\hat{\phi}
\tag{12.23}
$$

where $\hat{\boldsymbol{\eta}} = (\eta_1, \ldots, \eta_p)' = (X'X)^{-1}X'\mathbf{w}$. Note that

$$
\begin{bmatrix} \hat{\Delta}^* \\ \hat{\boldsymbol{\beta}}^* \end{bmatrix} = \begin{bmatrix} \hat{\Delta} \\ \hat{\boldsymbol{\beta}} \end{bmatrix}_{\text{update}}.
\tag{12.24}
$$

This update formula implies $\hat{\Delta}^* = \hat{\Delta} - \hat{\eta}_1\hat{\phi}$, which is equation (9) in Rosenbaum (1986). Note that Δ^* is the effect of dropping out of high school when the hidden covariate is included in the multiple regression model.

The sensitivity analysis is conducted by seeing if values for η and ϕ can be found which result in $\hat{\Delta}^*$ being small and/or changing sign from the sign of $\hat{\Delta}$. If such values on η and ϕ are also plausible, then the hypothesis test is inconclusive because it is excessively sensitive to a hidden covariate.

12.6.4.1 Selection of η and ϕ values

Values of the parameters η and ϕ at which to compute $\hat{\Delta}^*$ are selected under two assumptions. The first is that since matching was done on $\mathbf{x}$ without regard to u, the variability of $(u_1 - u_0)$ is not negligible. The second, most relevant to selection of η values, is that the hidden covariate, u, is completely unrelated to the observed covariates, $\mathbf{x}$, once assignment knowledge ('dropout' or 'stayer') is known. If these two assumptions are thrown out, there is no way to propose realistic possible values for η and ϕ.

Work on this topic is ongoing. For example, Rosenbaum (2010) shows that the most powerful test for detecting a treatment effect in a designed experiment is often not the most powerful test when applied to data from an observational study.

12.6.5 Sensitivity to hidden bias in group IDs

Say that the null hypothesis of the test is that the node G does not need to be in the model. Let $\mathbf{X}$ be the in-combination vector to a group ID. Let $\mathbf{Y}$ be the out-combination vector. Let $\boldsymbol{\beta}$ be the group ID's parameter vector and $P(\text{OGA} = oga) = h(\mathbf{X} = \mathbf{x}, \mathbf{Y} = \mathbf{y}|\boldsymbol{\beta})$ be the conditional probability mass function of the group ID's `Overall Goal Attainment` node when there is no hidden covariate. The expected value of OGA will be maximized for one particular out-combination, $\mathbf{y}$, and will hence be selected by the group to be their reaction to $\mathbf{x}$.

Now say that a hidden binary-valued covariate, U, is directly affecting the OGA node. In other words, U is an additional input variable to the group ID at every time point:

$$P(\text{OGA} = oga) = h(\mathbf{X} = \mathbf{x}, \mathbf{Y} = \mathbf{y}, U = u|\boldsymbol{\beta}^*) \qquad (12.25)$$

where $\boldsymbol{\beta}^* = \{\boldsymbol{\beta}' \, \boldsymbol{\beta}_{OGA}^{*\prime}\}'$ is the ID's parameter vector that includes parameters needed to represent the effect of U on the conditional distributions that define the OGA node ($\boldsymbol{\beta}_{OGA}^*$).

Perform the following steps to assess the sensitivity of a significant test result for G to a hidden covariate, U:

1. Remove G from the full model, and create a data set for U by specifying a 0 or 1 at each time point of which the group ID produces an action. To explore how such an unobserved covariate could be driving a particular decision,

place a '1' value only at those time points at which the group ID produced this decision. Add this hypothetical data on U to the political–ecological data set. Fix a set of values for β^*_{OGA}.

2. Conduct the MC hypothesis test wherein the full model is defined in the previous step, and the reduced model has both G and U removed.

3. If a set of plausible values for U and β^*_{OGA} can be found for which the test statistic is significant, then the obtained significant test result from the actual political–ecological data set could have been due to the hidden effect, U, rather than the node being tested for, G. In this case, the hypothesis test result for G is inconclusive.

This sensitivity to hidden bias analysis can be conducted by running **id**'s `mc_hypothesis_test` relation (see Chapter 5) iteratively over many different settings of U data and values of β^*_{OGA}.

12.7   Conclusions

The reliability of a political–ecological system simulator needs to be assessed before its outputs are used to make ecosystem management decisions. Here, reliability is assessed through (a) the simulator's sensitivity to misspecified parameters, and (b) the simulator's one-step-ahead prediction performance. Algorithms to compute these measures have been given and exercised on the East African cheetah EMT simulator. Results indicate that this simulator does not have excessive sensitivity to parameter value misspecification, and has a one-step-ahead prediction error rate that is low enough to allow the simulator to be used to reliably manage East African cheetahs.

A hypothesis-testing procedure has also been described that allows hypothesis tests to be used to choose among competing ID architectures (nodes and nodal connectivity), and/or individual probability values within one or several of an ID's defining conditional distributions. Being able to compare competing models supports efforts to improve the simulator's construct validity.

Before the results of an MC hypothesis test are used to modify one of the simulator's IDs, a sensitivity to hidden bias analysis should be performed as outlined above.

12.8   Exercises

1. Assess the simulator's sensitivity to parameters that define the `Overall Goal Attainment` node within the rural resident group IDs. Do this by running the sensitivity analysis option of **id** and allowing only parameters of the OGA node to define the optimization search space.

2. Do prediction errors increase as the prediction time becomes farther removed from the most recent observation time? Assess this by setting the most recent observation time to January 1, 2002. Then, compute predictions forward to January 1, 2009. Find error rates within a moving window of five time points over this range of predictions. Do these rates increase over this time interval?

3. Test the hypothesis that an actor's relative power is not important to a rural resident or pastoralist. Do this by modifying these group IDs so that the conditional distributions of actor image nodes place .99 probability on the *parity* value for every actor. Use the MC hypothesis test qualifier in the **id** language to test this reduced model against the full model of Chapter 7.

4. Program either the energy balance climate model of Shell and Somerville (2005) or the atmosphere–land coupled climate model of Fan *et al.* (2004). Conduct a deterministic sensitivity analysis under a critic-supplied output of zero mean temperature increase 50 years into the future. Which parameters must stray outside their intervals of plausible values before the model generates this output?

5. For the climate model chosen in the previous problem, and Sea Surface Temperature data from the Global Observing Systems Information Center (GOSIC 2010), compute OSARMSPE values for the annual average temperature at Bermuda, and at Murmansk, over the years 1990 through 2010.

Part III

ASSESSMENT

13

Current capabilities and limitations of the politically realistic EMT

13.1 Introduction

An assessment is given in this chapter of the politically realistic EMT developed in this book. The layout of this chapter is as follows. Section 13.2 contains an account of the EMT's current capabilities for finding politically realistic ecosystem management plans that also reach conservation goals for that ecosystem. Such plans are referred to in this book as Most Practical Ecosystem Management Plans (MPEMPs) (see Chapter 4). Section 13.3 contains a description of the current limitations of the EMT that need to be addressed to make the EMT more effective at finding such plans. Changes in the politics of wildlife management and changes in how wildlife management professionals are trained are described in Section 13.4 that, if enacted, would enhance the EMT's effectiveness. The last section contains some anticipated consequences of using a politically realistic EMT.

13.2 Current capabilities of the EMT

The general purpose EMT developed in this book can help decision makers working under political constraints to manage an ecosystem that is being affected by

Improving Natural Resource Management: Ecological and Political Models   Timothy C. Haas
© 2011 John Wiley & Sons, Ltd

human activities across several countries. This aid comes in the form of software that can find a management plan that delivers desired ecosystem outcomes for the price of having to effect changes in group belief structures. The analyst can be assured that these needed changes in beliefs are the smallest changes possible that will still lead groups to change their actions enough to cause desired ecosystem outcomes.

Currently, the EMT is capable of:

1. Supporting the construction of a political–ecological database through its taxonomy of ecosystem-relevant group actions and protocol for collecting such data.

2. Statistically fitting its simulator to a political–ecological data set.

3. Finding the MPEMP using this fitted simulator.

4. Assessing the sensitivity of the fitted simulator to parameter set misspecification.

5. Assessing the predictive validity of the fitted simulator through an estimate of its one-step-ahead prediction error rate.

6. Being freely downloaded through the book's website so that a politically realistic EMT can be built for any species that is affected by the activities of residents of just one country or by the residents of each of several countries.

To illustrate the EMT's capabilities and to show feasibility, a politically realistic EMT has been developed in this book that could be used to manage cheetah populations in East Africa. Using a minimal political–ecological data set, this EMT's simulator returns a prediction error rate of 0.5 (see Chapter 12) – making it superior to blind guessing.

Because of these demonstrations, this book shows not only that model-based management of a political–ecological system is possible, but that there now exists a freely available, web-based suite of software tools for doing it.

13.3   Current limitations of the EMT

Limitations of the current version of the EMT include:

1. Group IDs do not learn as they interact with each other and with the ecosystem. This is a different activity than using consistency analysis to update the simulator's parameter values in light of new data. A cognitively plausible mechanism is needed within a group ID's operation that updates its own parameter values as it experiences the consequences of its actions through time. For example, such parameter value updating applied to the Chosen Target node would allow a group ID to gradually shift the likely target of an action. An essential aspect of the needed cognitive plausibility would be

for these updating operations to be in the perceived direction of furthering the group's goals.

2. The group ID memory mechanism described in Chapter 6 needs to be implemented in the East African cheetah EMT simulator.

3. In the current version of the group ID architecture, groups are defined for the entire country so that the district-level information in the actions history data set cannot be used to study district-level cause-and-effect relationships between group actions and the ecosystem. Group IDs need to be refined so that a `Region` node can be used to represent decision-making behavior at a local, district level.

4. A set of online tutorials is needed to walk a new user through the steps of setting up an EMT, and using **id** to make ecosystem management computations.

5. It is not known to this author if the delete-d jackknife estimator of Chapter 12 is consistent when applied to the multivariate probability distributions represented by the group IDs of Chapter 7 and the ecosystem ID of Chapter 8. This question needs to be settled analytically or at least numerically for situations in hand.

6. Other group decision-making models need to be evaluated for their possible inclusion in the EMT simulator. The CLARION model of cognition (see Naveh and Sun 2006) or the Causes–Desires–Intentions (CDI) agent formalism of Saadi and Sahnoun (2006) could replace the group IDs of Chapter 7. The CDI agent formalism extends the well-known agent formalism of Beliefs–Desires–Intentions (BDI) of Bratman *et al.* (1988). This extension is achieved by incorporating degrees of belief, desire, and intention – and by building a mechanism that allows emotions to reduce the attractiveness of certain decision options. The design of the EMT simulator allows such alternative models to be easily experimented with.

7. Likewise, other ecosystem dynamics models need to be evaluated for their possible inclusion in the EMT simulator. For example, Individual-Based Models (IBMs) (see Grimm and Railsback 2005) could replace the system of SDEs for wildlife abundance presented in Chapter 8.

13.4   Supporting the EMT in the real world

13.4.1   Political needs

Once constructed, a politically realistic EMT would need to be maintained on the Web by a consortium of habitat-hosting countries and other countries that underwrite monitoring and species-protection programs within those habitat-hosting countries. Wildlife abundance data indexed by location and time from these programs would need to be publicly available on the EMT's website.

Political actions would need to be freely reported so that the predictive performance and construct validity of the EMT simulator could be continually improved using the methods of Chapter 12. The simulator's current predictive performance would need to be easily accessible by any web user so that questions and criticisms of its reliability and hence the credibility of its computed MPEMPs could be quickly answered. Doing so would reduce the amount of doubt, suspicion, and skepticism that is associated with many science-based management recommendations.

The simulator is complex. Significant computing capability would need to always be available to the consortium that is maintaining the EMT. The source code of all software used in the EMT would need to be freely accessible so that a 'priesthood' of ecosystem assessors does not have a chance to develop.

13.4.2   Educational needs

A major roadblock to implementing a politically realistic EMT in the real world is the shortage of ecosystem managers with the necessary breadth and depth of knowledge that is needed to fully understand the EMT simulator and the inferential statistics used to fit its parameters and assess its reliability. A new breed of ecosystem manager is needed, whose training includes advanced coursework in applied probability, wildlife population dynamics, simulation-based statistical inference, and political science. Most environmental management programs are deficient in several of these areas. But if the ecosystem to be managed has multiple stakeholders, then clearly a manager will need to be competent in both politics and the workings of ecosystems. Specifically, managers will need to receive interdisciplinary training at the level of political theory that supports the decision-making model of Chapter 6, wildlife population dynamics that supports the model of Chapter 8, and the statistical inference methods of Chapters 11 and 12.

Such managers are needed by the year 2020. To make this new training program specific, Table 13.1 lists standard texts and collections of readings in these fields

Table 13.1 Fields where modern ecosystem managers need in-depth knowledge along with suggested texts.

Field	Text(s)
Probability	Paolella (2006) (includes exercises)
Statistics	Good (2005) (includes exercises)
Jackknife tests	Hilmer and Holt (2000), Service *et al.* (2006)
Population dynamics	Lande *et al.* (2003), Allen (2007) (includes exercises), Iacus (2008)
Programming	Sedgewick and Wayne (2007) (includes exercises)
Social science	Harrison (2006), Phan and Amblard (2007), Sun (2006)
Political psychology	Krosnick and Chiang (2009), Sears *et al.* (2003), Jost and Sidanius (2004), Cottam *et al.* (2004)

that such managers need to have read. For a book in the list that has exercises, performing several exercises given at the end of each chapter is a critical path to understanding that book's contents.

13.5   Consequences of using a politically realistic EMT

What if the MPEMP computed from a reliable simulator predicted that, even under this plan, the managed species would still go extinct? Such a result would add credence to the view that to avoid the extinction of a species, only three options would remain: captive populations, translocated populations, and large-scale separation of human and wildlife populations.

Bashir *et al.* (2004) note that captive cheetah populations are not self-sustaining and hence are regularly supplemented by importing animals from the wild population. So, for at least the cheetah, captive breeding may not be a solution. The second solution, that of developing an 'Ark' population in another location that does not have the insurmountable problems of the current habitat-hosting countries, has been studied by Donlan (2005). In this article, Donlan proposes the establishment of wild populations of threatened African species in the southwest area of the USA. One form of the third solution involves the creation of protected wildlife 'corridors' that run along the boundaries between countries (see Metcalfe and Kepe 2008). These transboundary corridors would allow animal migrations and would mitigate boundary disputes. Another form of this third solution is a large increase in the size of existing reserves coupled with the creation of new reserve areas. All of these reserves would need regular anti-poaching patrols.

13.5.1   One solution to the problem posed in Chapter 1

Humankind needs to be able to discuss in specific terms what impact its activities have on the chances that a species will become extinct. A politically realistic EMT allows the computation of one way to assess such a risk: a species is at risk of going extinct if the probability that there are fewer than 10 animals alive 50 animal-generations into the future is greater than .01 (see Chapter 1).

Many messages in the popular press concerning the plight of wildlife contain little detail and are wrapped in hype, spin, and/or dramatic portrayals of wildlife in distress – all with the intention of inspiring sympathy. This author believes that individuals in media-rich countries, however, often ignore such messages when they are not congruent with their existing belief structures because such messages do not contain enough reliable information to challenge their existing understanding of the world. The EMT of this book takes a different approach: present an up front, detailed, and open picture of the state of a species accompanied with accepted measures of the reliability of this picture – and then propose politically realistic solutions to the management problem. Having such a picture and attendant management proposals may cause individuals who are otherwise skeptical to internalize the problem. To this author, it seems reasonable to think that such internalization would increase the likelihood of such skeptical individuals participating in humankind's ecosystem management decision making.

Such participation, should it occur, is seen by this author as desirable because, from now on, humankind is the keeper of all wildlife on earth and will decide either explicitly or as a by-product of unintentioned actions whether these animals and plants persist or perish.

APPENDICES

Appendix A

Heuristics used to assign hypothesis values to parameters

Heuristics based on theories of political decision making (see Chapter 6) are used to assign hypothesis values to group ID conditional distribution parameters. These heuristics are given here.

President IDs

Economic Resources Change: Table 7.1, Chapter 7, gives the values for the represented actions. Change occurs only if Subject equals *self*. Poaching hurts the tourism industry run by campaign donors so this input action causes a decrease in both immediate and future economic resources.

Economic Resources: Chance of increased economic resources increases with increased level of economic resources one time step back and/or positive Economic Resources Change values.

Use of Military Force: Values are taken from Table 7.1, Chapter 7.

Military Resources Change: See Table 7.1, Chapter 7.

Military Resources: Chance of increased military resources increases with increased level of military resources one time step back and/or positive Military Resources Change values.

Actor Affect, Actor Relative Power: Both rural residents and pastoralists are neutral and less powerful actors.

Improving Natural Resource Management: Ecological and Political Models   Timothy C. Haas
© 2011 John Wiley & Sons, Ltd

Campaign Donor Audience Change: Chance of decreased donor satisfaction is caused by negative economic change. Donors are mostly concerned with economic changes.

Campaign Donor Audience Satisfaction: A memory of donor dissatisfaction and/or negative donor change increases the chance of dissatisfied donors.

Military Audience Change: A militaristic action by an enemy causes a negative change in the military's attitude.

Military Audience Satisfaction: Similar to Campaign Donors.

Maintain Order Goal: Chance of being attained increases with increasing military resources.

Maintain Political Power Goal: Chance of attainment increases when both donors and the military are satisfied.

Target Affect, Target Relative Power: See Actor Affect, Actor Relative Power.

Scenario Economic Resources Change: *create_wildlife_reserve* and *open_wildlife_reserve* are seen to stimulate the economy; *request_increased_antipoaching* and *suppress_riot* are seen to cost money. See Table 7.1, Chapter 7.

Scenario Use of Military Force: See Table 7.1, Chapter 7.

Scenario Military Resources Change: Only *request_increased_antipoaching* and *suppress_riot* are seen to require military resources. See Table 7.1, Chapter 7.

Scenario Military Resources: See Military Resources.

Scenario Campaign Donor Audience Change: See Campaign Donor Audience Change.

Scenario Military Audience Change: The military want to see strong reactions to external threats, that is, a militaristic response to militaristic actions of enemies.

Scenario Economic Resources: See Economic Resources.

Scenario Campaign Donor Audience Satisfaction: See Campaign Donor Audience Satisfaction.

Scenario Military Audience Satisfaction: See Military Audience Satisfaction.

Scenario Maintain Order Goal: A militarily effective action when Maintain Order Goal is unattained causes this goal to be attained. If Maintain Order Goal is attained, a militarily effective action is perceived to backfire as unnecessary force and cause militaristic reactions. Hence, in this case, this goal is unattained.

Scenario Maintain Political Power Goal: See Maintain Political Power Goal.

Scenario Overall Goal Attainment: Maintaining political power is more important than maintaining order.

EPA IDs

Cheetah, Herbivore Fraction Detected: Fraction of the region's area on which cheetahs and herbivores are detected. Read from bulletin board postings made by the ecosystem ID.

Cheetah Prevalence, Herbivore Prevalence: Prevalence is expressed on a scale of 0 to 1 by making prevalence the dependent variable in a logit model. This logit model possesses one predictor variable, the associated Fraction Detected node.

Rural Resident Poaching Rate: Rural resident poaching actions increase belief in rural residents frequently poaching.

Pastoralist Poaching Rate: Pastoralist poaching actions increase belief in pastoralists frequently poaching.

Overall Poaching Rate: Belief in frequent poaching increases as either rural resident or pastoralist poaching rates increase.

Actor Affect, Actor Relative Power: The president is a neutral but more powerful actor. Rural residents and pastoralists are neutral and less powerful actors.

Economic Resources Change: The EPA's economic resources reside in its operating budget which is set by the president. It is believed that a presidential order for more EPA activity will be accompanied by more resources.

Economic Resources: Same as president IDs.

President Audience Change: The president has little interest in biodiversity protection but is very interested in promoting economic opportunities for the rural residents. Therefore, the president is seen to respond negatively only to heavy poaching and riots. A positive response is seen to the input actions of *clear new land* and the president's own actions of *open a reserve*, *create a reserve*, or *request increased antipoaching enforcement*.

President Audience Satisfaction: A memory of presidential dissatisfaction with decreased presidential satisfaction causes a high chance of presidential dissatisfaction.

Protect Ecosystem Goal: Low values of either Cheetah Prevalence or Herbivore Prevalence and high values of Poaching Rate cause a high chance of this goal being unattained.

Expand Budget Goal: A satisfied president increases the chance of an increased budget.

Target Affect, Target Relative Power: See Actor Affect, Actor Relative Power.

Scenario Rural Resident Poaching Rate, Scenario Pastoralist Poaching Rate: If situation rates are believed to be high and nothing is done about it in the scenario, these nodes inherit the associated situation node beliefs.

Scenario Cheetah and Herbivore Prevalence, Scenario Poaching Rate: Increased anti-poaching enforcement is believed to yield modest gains in protecting biodiversity. Habitat loss due to the clearing of new land is not incorporated in this version of the model.

Scenario Economic Change: A negative ecosystem report is seen to cause a small increase in resources to the EPA. All other actions are believed to cause no change in resources. Decreased Situation Economic Change causes a negative economic change in the scenario.

Scenario Economic Resources: Same as president IDs.

Scenario President Audience Change: See President Audience Change.

Scenario President Audience Satisfaction: See President Audience Satisfaction.

Scenario Protect Ecosystem Goal: See Protect Ecosystem Goal.

Scenario Expand Budget Goal: See Expand Budget Goal.

Scenario Overall Goal Attainment: Survival and growth of the agency (expanding the budget) is as important as protecting the ecosystem.

Rural resident IDs

Cheetah, Herbivore Fraction Detected: Same as EPA IDs.

Cheetah Prevalence, Herbivore Prevalence: Same as EPA IDs.

Actor Affect, Actor Relative Power: The president and EPA are both neutral but more powerful actors.

Economic Resources Change: Increased anti-poaching enforcement causes a high chance of decreased economic resources.

Economic Resources: Same as president IDs.

Family Audience Change: Opening a wildlife reserve causes a high chance of increased family satisfaction.

Family Audience Satisfaction: A memory of family dissatisfaction with decreased family satisfaction causes a high chance of family dissatisfaction.

Family Satisfaction Goal: Family dissatisfaction causes a high chance of a dissatisfied family goal.

Avoid Prosecution Goal: Increased anti-poaching enforcement causes higher chances of being prosecuted, hence this goal being unattained.

Target Affect, Target Relative Power: See Actor Affect, Actor Relative Power.

Scenario Economic Resources Change: Increased poaching increases the chance of a positive economic resource change.

Scenario Economic Resources: Same as president IDs.

Scenario Family Audience Change: Positive economic change increases the chance of increased family satisfaction.

Scenario Family Audience Satisfaction: Increased family satisfaction causes a high chance of a satisfied family.

Scenario Family Satisfaction Goal: A satisfied family causes a high chance of the family goal being satisfied.

Scenario Avoid Prosecution Goal: Increased poaching increases the chance of this goal being unattained.

Scenario Overall Goal Attainment: The family goal is more important than avoiding prosecution.

Pastoralist IDs

Cheetah Fraction Detected: Same as EPA IDs.

Cheetah Prevalence: Same as EPA IDs.

Pastoralist Territory Size: Taken from GIS maps of pastoralist range by country.

Actor Affect, Actor Relative Power: The president and the EPA are both neutral but more powerful actors.

Livestock Change: Increasing anti-poaching enforcement threatens livestock.

Livestock Resources: Similar to the president ID's node, Economic Resources.

Family Audience Change: Decreased family satisfaction is likely when a wildlife reserve is created or when anti-poaching enforcement is increased.

Family Audience Satisfaction: A memory of family dissatisfaction with decreased family satisfaction causes a high chance of family dissatisfaction.

Family Satisfaction Goal: Family dissatisfaction causes a high chance of this goal being unattained.

Protect Livestock Goal: Decreased livestock increases the chance of this goal being unattained.

Avoid Prosecution Goal: Same as rural resident IDs.

Target Affect, Target Relative Power: See Actor Affect, Actor Relative Power.

Scenario Pastoralist Territory Size: Size of pastoralist range caused by the pastoralists' action. Effective lobbying is seen as increasing this range size.

Scenario Livestock Resources Change: Pastoralists believe that, by poaching, they are protesting government actions of reserve creation and the placement of restrictions on hunting. Pastoralists believe that such forms of political protest will cause a positive livestock resources change.

Scenario Livestock Resources: Similar to the president ID's node, Scenario Economic Resources.

Scenario Family Audience Change: Same as rural resident IDs.

Scenario Family Audience Satisfaction: Same as rural resident IDs.

Scenario Family Satisfaction Goal: Same as rural resident IDs.

Scenario Protect Livestock Goal: Adequate Scenario Livestock Resources raises the chance of this goal being attained.

Scenario Avoid Prosecution Goal: Increased poaching increases the chance of this goal being unattained.

Scenario Overall Goal Attainment: Protecting livestock is more important than avoiding prosecution.

Conservation NGOs ID

Input Action: A conservation NGO listens for input actions that affect (a) large areas of wildlife habitat or large numbers of wildlife, and (b) proposals for such actions.

(Country) Wildlife Prevalence: A conservation NGO is more sensitive to cheetah losses than herbivore losses.

Donors Change: External donors are very displeased with failure to reach wildlife conservation goals.

Cheetah Density, Herbivore Density: Read from the posted cheetah detection and herbivore abundance marginal distributions generated by the ecosystem ID.

Cheetah Prevalence, Herbivore Prevalence: Logit models in the density nodes with positive abundance thresholds.

(Country) Rural Resident Poaching Rate: Rural resident poaching actions increase belief that rural residents frequently poach. (Country) is Kenya, Tanzania, or Uganda.

(Country) Pastoralist Poaching Rate: Pastoralist poaching actions increase belief in pastoralists frequently poaching.

Rural Resident Poaching Rate: Belief in frequent rural resident poaching increases as rural resident poaching occurs in at least one of the three countries.

Pastoralist Poaching Rate: Belief in frequent pastoralist poaching increases as pastoralist poaching occurs in at least one of the three countries.

Economic Resources Change: A conservation NGO's economic resources reside in its operating budget which is influenced by the satisfaction of external donors.

Economic Resources: Same as president IDs.

(Country) President Audience Change: Same as EPA IDs.

(Country) President Audience Satisfaction: Same as EPA IDs.

External Donors Audience Change: External donors react negatively to poaching actions.

External Donors Audience Satisfaction: Negative changes cause high chance of donors being dissatisfied.

Conserve Wildlife Goal: Low values of Cheetah Prevalence or Herbivore Prevalence and high values of Poaching Rate cause a high chance of goal being unattained.

Maintain Relations Goal: One or more dissatisfied presidents causes this goal to be unattained.

Expand Budget Goal: More economic resources cause a high chance that this goal is attained.

Scenario Poaching Rate: If the situation of rural resident poaching rates and/or pastoralist poaching rates is high and nothing is done about it in the scenario, this node inherits the situation beliefs.

Scenario Cheetah and Herbivore Prevalence: Increased anti-poaching enforcement is believed to yield modest species abundance increases.

Scenario Economic Change: Same as president IDs.

Scenario Economic Resources: Same as president IDs.

Scenario (Country) President Audience Change: See (Country) President Audience Change.

Scenario (Country) President Audience Satisfaction: See (Country) President Audience Satisfaction.

Scenario External Donors Audience Change: Animal translocation and education initiatives cause positive external donor changes.

Scenario External Donors Audience Satisfaction: See External Donors Audience Satisfaction.

Scenario Conserve Wildlife Goal: See Conserve Wildlife Goal.

Scenario Maintain Relations Goal: See Maintain Relations Goal.

Scenario Expand Budget Goal: See Expand Budget Goal.

Scenario Overall Goal Attainment: Survival and growth of a conservation NGO (expanding its budget) is as important as conserving wildlife.

Appendix B

Cluster computing version of Hooke and Jeeves search

The Hooke and Jeeves optimization algorithm can be modified to run on a cluster computing system. A parallel version of this algorithm has been developed, called Multiple Dimensions Ahead Search (MDAS). This algorithm simultaneously searches the next M variables for a minimum.

For reference, the classic Hooke and Jeeves algorithm (Hooke and Jeeves 1961) as presented by Kaupe (1963) with corrections by Bell and Pike (1966) and Tomlin and Smith (1969) is as follows.

```
procedure direct search (psi, K, Spsi, DELTA, rho, delta, S, converge,
                         maxeval);

value         K, DELTA, rho, delta, maxeval;
integer       K, maxeval;
array         psi;
real          DELTA, rho, delta, Spsi;
real procedure S;
boolean       converge;

begin;
integer k, eval;
array    phi[1:K], s[1:K];
real     Sphi, SS, theta;

for k = 1 step 1 until K do s[k] = DELTA;
Spsi = S(psi);
eval = 1;
converge = true;
1: SS = Spsi;
```

```
for  k = 1 step 1 until K do phi[k] = psi[k];
call procedure E;
if SS < Spsi then begin
   2: for k = 1 step 1 until K do
      if phi[k] > psi[k] AND psi[k] = s[k] AND s[k] < 0 then s[k] = -s[k];
      theta = psi[k];
      psi[k] = phi[k];
      phi[k] = 2 * phi[k] - theta;
   end
   Spsi = SS;
   if eval >= maxeval then begin
      3: converge = false;
      go to EXIT;
   end;
   SS = Sphi = S(phi);
   eval = eval + 1;
   call procedure E;
   if SS >= Spsi then go to 1;
   for k = 1 step 1 until K do
      if abs(phi[k] - psi[k]) > .5 * abs(s[k]) then go to 2;
   end
end
if DELTA >= delta then begin
   if eval > maxeval then go to 3;
   else DELTA = rho * DELTA;
   for k = 1 step 1 until K do s[k] = rho * s[k];
   go to 1;
end;
EXIT:
end direct search;

procedure E;
for k = 1 step 1 until K do
   phi[k] = phi[k] + s[k];
   Sphi = S(phi);
   eval = eval + 1;
   if Sphi < SS then SS = Sphi;
   else begin
      s[k] = -s[k];
      phi[k] = phi[k] + 2 * s[k];
      Sphi = S(phi);
      eval= eval + 1;
      if Sphi < SS then SS = Sphi;
      else phi[k] = phi[k] - s[k]
   end
end
end E;
```

Modification for stochastic objective functions

Before describing cluster computer implementations of the Hooke and Jeeves algorithm, one important modification of it needs to be described. The g_{CA} objective function that constitutes the **Maximize** step of consistency analysis actually produces deviates from a random variable, G_{CA}, because the Hellinger distances that define g_{CA} are replaced in practice with estimates of these distances. Therefore, as the Hooke and Jeeves algorithm executes, a test for whether the objective function has become smaller when moving to a new point in the solution space needs to acknowledge the variability in objective function values that is due to the stochasticity of G_{CA} as separate from the variability in g_{CA} values that are due to its being evaluated at different points in the solution space.

Alkhamis and Ahmed (2006) give one way to adjust for this second source of variability as follows. Say that the objective function is the average of N deviates from a random variable, Y, but the desired function to minimize is $E[Y]$. Say that the algorithm has moved to the point x_0 and is ready to test whether the objective function is smaller at a new point x_1. The test 'Is $\bar{y}(x_1) < \bar{y}(x_0)$?' ignores the stochastic component of $\bar{Y}(.)$'s variability and hence is replaced with the test 'Is

$$\bar{y}(x_1) < \bar{y}(x_0) - t_{1-\alpha/2,N-1}\hat{\sigma}_Y/\sqrt{N}?'$$

In this expression, $t_{1-\alpha/2,N-1}$ is the $1 - \alpha/2$ quantile of the t distribution with $N - 1$ degrees of freedom.

The adjustment of Alkhamis and Ahmed (2006) cannot be used when the Hooke and Jeeves algorithm is applied to the objective function g_{CA}. This is because g_{CA} is not a simple average of N deviates. Therefore, the adjustment used in **id** is as follows. At the beginning of a **Maximize** run, recompute g_{CA} n_0 times and compute the sample standard deviation of these values, $s_{g_{CA}}$. Then, inside the Hooke and Jeeves algorithm, replace all occurrences of 'if $(-g_{CA}(x_1) < -g_{CA}(x_0))$' with 'if $(-g_{CA}(x_1) < -g_{CA}(x_0) - Cs_{g_{CA}})$' where C is a constant to control the false positive rate. In the runs reported in Chapter 11, $C = 1.5$.

MDAS description

As in the pseudo code above, let K be the number of independent variables. For $M = 3$, exhaustively evaluate all possible visited locations for the next three dimensions in the 'for' loop of procedure E, above. This requires $2 + (3 * 2) + (3 * 3 * 2) = 3^3 - 1 = 26$ parallel evaluations of the objective function, S ( ). If an improvement is found at a particular set of dimensions, search these same three dimensions again to see if a further improvement can be had. Continue this way until these three dimensions do not yield an improvement. When this happens, move to the next three dimensions.

When there are at least 26 child processors available, this procedure results in a six times speed-up over the worst case during Hooke and Jeeves search when K is a multiple of three. Note that the sequential version of the 'for' loop in procedure E will perform up to $2K$ function evaluations.

After the child processors return their function evaluation values, the master processor checks these values for a new minimum. If found, the master processor stores this new best solution in the array psi. Therefore, the master processor performs this check every W seconds where W is the wall-clock time needed to compute the objective function on a single processor. This characteristic of the algorithm is important because it makes the algorithm robust to the loss of any part of the distributed computing system.

Hooke and Jeeves search over the variables in procedure E is a form of greedy search. This same search in MDAS is exhaustive in each set of three dimensions and hence should be able to find lower minima than Hooke and Jeeves search.

Also note that a parallel direct search algorithm applied to a constrained optimization problem that possesses some mixture of nonlinear and implicit constraints has the following advantages:

1. Gradients are not computed so that a step into an infeasible location cannot happen.

2. Proofs exist that the Hooke and Jeeves algorithm is guaranteed to converge, see Lewis *et al.* (2000).

3. As long as 26 processors can be employed, a six-times speed-up over the worst case of Hooke and Jeeves search is guaranteed.

4. Hooke and Jeeves search can accommodate ordinally valued, discrete variables, a fact seldom mentioned in the optimization literature.

5. Parallelism is not achieved by dividing the region into subregions and then doing a parallel search of these subregions as is done with Simplex methods (see, e.g., Lewis *et al.* 2000) or simple box-subregion methods (e.g. see Leitão and Schiozer (1999) for a search algorithm that employs parallel Hooke and Jeeves searches each of which is started at a different point in the parameter space). Such subregions can be infeasible and very far from a solution point. For the parallel Simplex algorithm in particular, its stability has been studied only in the unconstrained case (Lewis *et al.* 2000) – and these authors warn users of possible instability of the algorithm when constraints are present.

6. In general, to produce a $2M$ speed-up over worst case sequential Hooke and Jeeves, MDAS needs to be run on a cluster computer having $3^M - 1$ processors. For example, if $M = 5$ (one order of magnitude speed-up), a 242-processor cluster computer is needed. With present supercomputer technology, a 16 times speed-up is about the upper limit of feasible speed-up as this would require 6560 processors ($M = 8$).

With only 8 processors and $M = 2$, the MDAS algorithm achieves a four times speed-up over the worst case sequential Hooke and Jeeves. Many wildlife management agencies could afford such a cluster of computers but would not be able to afford time on a (say) 1300-processor supercomputer.

What is the penalty in terms of computational speed-up that this lack of funds entails? Say that there are 50 users of the supercomputer. Then, any one user in effect has a 26-processor computer and hence can run six times faster versus four times faster using an in-house cluster computer. This difference is not enough to allow significantly more complicated models to be fitted. And, if uncertainties in support for purchasing supercomputer time are present, the in-house cluster computer option may be more attractive.

Parallel Best Step Search (PBSS)

A different approach to parallelizing the Hooke and Jeeves algorithm is to find the best single-step move by computing all possible single steps from the current point in parallel. Then, repeat this single-step activity until the function can no longer be reduced. This is accomplished with the following modification to procedure E:

```
procedure E;
Sphi = SS;
start = true;
if SS - Sphi > 1.e-6 or start = true then begin
   start = false;
   SS = Sphi;
   j = 0;
   for k = 1 step 1 until K do
      for i = 1 step 1 until K do
         phitry[i] = phi[i];
      end
      j = j + 1;
      phitry[j][k] = phi[k] + s[k];
      jval[j] = j;
      j = j + 1;
      phitry[j][k] = phi[k] - s[k];
      jval[j] = j;
   end
   for k = 1 step 1 until j do (parallelize)
      stry[k] = S(phitry[k]);
   end
   eval = eval + j;
   sort(stry, carry along jval);
   Sphi = stry[jval[1]];
   if Sphi > SS end E;
   for k = 1 step 1 until K do
      phi[k] = phitry[jval[1]][k];
   end
   SS = Sphi;
end
end E;
```

If there are K processors, PBSS will find the best single-coordinate move in parallel. PBSS will step the objective function down one step towards its minimum every W seconds. The fastest that sequential Hooke and Jeeves can achieve is one step down every W seconds, but only if every left-step trial function evaluation on every coordinate is successful – not likely on many functions.

PBSS, then, is appropriate on cluster computers with as many processors as variables in the optimization problem. This would typically be a shared cluster computer environment. The comments about how a wildlife conservation agency would be able to budget such computer time discussed above in relation to MDAS apply here to PBSS.

Parallel Search Over Subsets of Variables (PSOSV)

There is an algorithm that is a compromise of PBSS and sequential Hooke and Jeeves, called here *Parallel Search Over Subsets of Variables* (PSOSV). Letting L be the number of child processors, PSOSV is defined as follows:

1. Run the main program of sequential Hooke and Jeeves on a parent processor. Run a parallel version of procedure E, above, called procedure PE. Procedure PE consists of the following activities:

   (a) Partition the vector of variables into L subvectors. Give each subvector and the starting point to a child processor and have it run a sequential Hooke and Jeeves procedure E up to either the end of its subvector or the first move that produces an objective function reduction. Child processors check the JavaSpaces bulletin board for abort messages after each function evaluation.

   (b) If a processor finds a point of objective function reduction, that processor posts the associated point and a message that it found a point of objective function reduction.

   (c) The parent processor reads this message and posts a message to all processors to (i) stop their searches, and (ii) restart execution of procedure E at this new point and at the beginning of their subvector of variables.

   (d) The previous two steps are iterated until all child processors reach the end of their subvectors and no child has found another point of objective function reduction. When this happens, procedure PE returns to the main program of the sequential Hooke and Jeeves algorithm running on the parent processor.

2. Once back in the main program, the step size is reduced and procedure PE is called again.

This algorithm combines the opportunistic search behavior of the sequential Hooke and Jeeves procedure E with simultaneous search of portions of the vector of variables. Hence, this algorithm scales well across different values of L making it a practical search algorithm for a distributed cluster of workstations in and across branch offices of a wildlife conservation agency.

Recommended computer hardware upgrades

In this author's opinion, coarse-grain parallel algorithms are currently the least expensive and most efficient approaches to solving an optimization problem in which the objective function consists of a Monte Carlo simulation. This view is in agreement with studies of parallel quasi-Newton optimization algorithms conducted by Eldred and Schimel (1999). This is due in part to the steady cost reduction of multicore PCs, in conjunction with the typically large expense of increasing processor-to-processor message and/or data passing.

Therefore, for a wildlife conservation agency to inexpensively and efficiently use **id** to maintain an EMT, it is recommended that desktop computers be replaced with PCs that have at least octal-core processors. Each such PC should be on the Internet and be left on at night. Then, an analyst could run his/her PC as the parent processor in the PSOSV algorithm above, and either estimate the simulator's parameters with consistency analysis or perform a deterministic sensitivity analysis using the search algorithm of Chapter 12. These optimization runs would be efficient because there would not be massive amounts of messages or data passing in and out of the JavaSpace – and each PC would perform (on average) several objective function evaluations in which Monte Carlo simulations are performed in parallel across multiple cores via Java's multithreading capability. This ability to perform Monte Carlo simulations in parallel is crucial because the objective function's most expensive component is the Monte Carlo solution to the system of stochastic differential equations inside the ecosystem ID.

Solution times can be cut by two or more orders of magnitude if time on large cluster computers can be purchased. At first glance, this approach appears to be the optimal way to estimate parameters and study the sensitivity of an EMT simulator. A fundamental goal of this book, however, has been to create an EMT that is open and not dependent on funding from any one source. Relying exclusively on continued support to purchase supercomputer time would make an EMT completely dependent on the source of computing time funds. Therefore, it is recommended to use supercomputer time opportunistically, that is, to speed up (say) the initial parameter estimation computation and/or the initial sensitivity analysis. Thereafter, the EMT simulator could be maintained on the supercomputer until and if funding for supercomputer time runs out. From that point on, the EMT simulator would be maintained via a JavaSpaces cluster computing procedure running exclusively on in-house office computers.

References

Aarts, E. and Korst, J. (1989) *Simulated Annealing and Boltzmann Machines,* John Wiley & Sons, Inc., New York.

Agresti, A. (2002) *Categorical Data Analysis,* Second Edition, John Wiley & Sons, Inc., New York.

Alkhamis, T. M. and Ahmed, M. A. (2006) A modified Hooke and Jeeves algorithm with likelihood ratio performance extrapolation for simulation optimization, *European Journal of Operational Research,* 174, pp. 1802–1815.

Allen, E. (2007) *Modeling with Itô Stochastic Differential Equations,* Springer-Verlag, New York.

Allen, K. R. (1963) Analysis of stock-recruitment relations in Antarctic fin whales, *Conseil International pour l'Exploration de la Mer – Rapport et Proces Verabaux,* 164, pp. 132–137.

Anderies, J. M. (2002) The Transition from Local to Global Dynamics: a Proposed Framework for Agent-Based Thinking in Social-Ecological Systems, in M. A. Janssen (Ed.), *Complexity and Ecosystem Management: the Theory and Practice of Multi-Agent Systems,* Edward Elgar Publishing, New York, pp. 13–34.

Arkive (2009) Blue whale *(Balaenoptera musculus).* Retrieved August 27, 2009, from `www.arkive.org/blue-whale/balaenoptera-musculus/info.html`

Armbruster, P., Fernando, P., and Lande, R. (1999) Time frames for population viability analysis of species with long generations: an example with Asian elephants, *Animal Conservation,* 2, pp. 69–73.

Australian Whale Conservation Society, (2010) Whaling today. Retrieved May 29, 2010, from `www.awcs.org.au` (click `Whaling today`).

Baars, B. J. (1988) *A Cognitive Theory of Consciousness,* Cambridge University Press, Cambridge.

Babbie, E. (1992) *The Practice of Social Research,* Sixth Edition, Wadsworth, Belmont, California.

Improving Natural Resource Management: Ecological and Political Models Timothy C. Haas
© 2011 John Wiley & Sons, Ltd

Baldauf, S. (2008) How Kenya came undone, *The Christian Science Monitor,* January 29. Retrieved May 29, 2010, from www.csmonitor.com/World/Africa/2008/0129/p01s04-woaf.html

Barber-Meyer, S. M., Kooyman, G. L., and Ponganis, P. J. (2007) Estimating the relative abundance of emperor penguins at inaccessible colonies using satellite imagery, *Polar Biology,* 30, pp. 1565–1570.

Bashir, S., Daly, B., Durant, S. M., Förster, H., Grsham, J., Marker, L., Wilson, K., and Friedmann, Y. (Eds.) (2004) Global Cheetah (*Acinonyx jubatus*) Monitoring Workshop, Final Workshop Report, Conservation Breeding Specialist Group (SSC/IUCN), Endangered Wildlife Trust (2004). Retrieved May 29, 2010, from www.catsg.org/cheetah/20_cc-compendium/home/index_en.htm (click Library and then Reports).

Bell, M. and Pike, M. C. (1966) Remark on algorithm 178 [E4] Direct Search, *Communications of the ACM,* 9, pp. 684–685.

BitLaw (2009a) Works unprotected by copyright law. Retrieved May 29, 2010, from www.bitlaw.com/copyright/unprotected.html

——— (2009b) Database legal protection. Retrieved May 29, 2010, from www.bitlaw.com/copyright/unprotected.html

Brandt, M. W. and Santa-Clara, P. (2002) Simulated likelihood estimation of diffusions with an application to exchange rate dynamics in incomplete markets, *Journal of Financial Economics,* 63, pp. 161–210.

Bratman, M. E., Israel, D. J., and Pollack, M. E.(1988) Plans and resource-bounded practical reasoning, *Computational Intelligence,* 4 (4), pp. 349–355.

Brewer, G. and de Leon, P. (1983) *Foundations of Policy Analysis,* Dorsey Press, Homewood, Illinois.

British Broadcasting Corporation News (2006) The forces that drive Japanese whaling, *British Broadcasting Corporation News,* June 15. Retrieved May 29, 2010, from http://news.bbc.co.uk/2/hi/5080508.stm

Brock, W. and Xepapadeas, A. (2002) Optimal ecosystem management when species compete for limiting resources, *Journal of Environmental Economics and Management,* 44 (2), pp. 189–220.

Buckland, S. T., Anderson, D. R., Burnham, K. P., Laake, J. L., Borchers, D. L., and Thomas, L. (2001) *Introduction to Distance Sampling,* Oxford University Press, Oxford.

Carmines, E. G. and Zeller, R. A. (1979) *Reliability and Validity Assessment,* Sage, Beverly Hills, California.

Cat Specialist Group (2007) *Acinonyx jubatus,* in IUCN 2007, *The IUCN Red List of Threatened Species.* Retrieved September 1, 2008, from www.iucnredlist.org

Chajewska, U., Koller, D., and Ormoneit, D. (2001) Learning an agent's utility function by observing behavior, *18th International Conference on Machine Learning (ICML'01),* pp. 35–42. Retrieved May 29, 2010, from http://ai.stanford.edu/~koller/Papers/Chajewska+al:ICML01.pdf

Chazan, N., Lewis, P., Mortimer, R. A., Rothchild, D., and Stedman, S. J. (1999) *Politics and Society in Contemporary Africa,* Lynne Rienner, Boulder, Colorado.

CIA (2010a) *The World Factbook – Kenya.* Retrieved October 2, 2010, from https://www.cia.gov (click Library > Publications > The World Factbook and then select Kenya from the select a country menu).

_____ (2010b) *The World Factbook – Tanzania.* Retrieved October 2, 2010, from
https://www.cia.gov (click Library > Publications > The World
Factbook and then select Tanzania from the select a country menu).

_____ (2010c) *The World Factbook – Uganda.* Retrieved October 2, 2010, from
https://www.cia.gov (click Library > Publications > The World
Factbook and then select Uganda from the select a country menu).

CITES (2010) *Appendix I: Endangered Species.* Retrieved February 24, 2010, from
http://www.cites.org/eng/app/appendices.shtml

Clark, C. W. (1976) A delayed-recruitment model of population dynamics, with an application to baleen whale populations, *Journal of Mathematical Biology,* 3, pp. 381–391.

Coase, R. H. (1937) The nature of the firm, *Economica,* 4, pp. 386–405.

_____ (1960) The problem of social cost, *Journal of Law and Economics,* 3, pp. 1–44.

_____ (1984) The new institutional economics, *Journal of Institutional and Theoretical Economics,* 140, pp. 229–31.

Cochran, W. G. (1965) The planning of observational studies of human populations (with discussion), *Journal of the Royal Statistical Society,* Series A, 128, pp. 134–55.

Cooper, G. F. (1987) Probabilistic inference using belief networks is NP hard, *Research Report KSL-87-27,* Medical Computer Science Group, Stanford University.

Cottam, M. L., Dietz-Uhler, B., Mastors, E., and Preston, T. (2004) *Introduction to Political Psychology,* Psychology Press (Taylor & Francis), Abingdon.

Demetriou, D. (2008) Japanese whale industry's flagship restaurant to close, *Telegraph,* November 13. Retrieved December 30, 2009, from www.telegraph.co.uk

Donlan, J. (2005) Re-wilding North America, *Nature,* 436, pp. 913–914.

Duffy, R. (2000) *Killing for Conservation,* Indiana University Press, Bloomington, Indiana.

Eldred, M. S. and Schimel, B. D. (1999) Extended parallelism models for optimization on massively parallel computers, *Third World Congress of Structural and Multidisciplinary Optimization,* Amherst, New York, May 17–21. Available as a *Sandia Laboratories Report SAND99-1295C,* Albuquerque, New Mexico.

Efron, B. (1983) Estimating the error rate of a prediction rule: improvement on cross-validation, *Journal of the American Statistical Association,* 78 (382), pp. 316–331.

eftec (2009) Economics of subsidies to whaling, *Economics for the Environment Consultancy (eftec),* Report prepared for WWF-UK and WDCS, main contributors Dr. Rob Tinch and Zara Phang (2009). Retrieved January 1, 2010, from www.panda.org/iwc (click economics of subsidies to whaling pdf).

EISA (2010) *Tanzania: Fact File.* Retrieved May 4, 2010, from www.eisa.org
.za/WEP/tan7.htm

Eisenhower, D. D. (1961) Farewell address, delivered January 17, 1961, *The American Presidency Project,* J. T. Woolley and G. Peters (Eds.), University of California at Santa Barbara, Santa Barbara, California. Retrieved February 26, 2010, from http://www
.presidency.ucsb.edu/ws/?pid=12086

eMap-International (2010) Retrieved January 13, 2010, from www.emap-int.com

Espeland, R. H. (2007) When neighbours become killers: ethnic conflict and communal violence in Western Uganda, *CMI Working Paper,* 2007: 3, Chr. Michelsen Institute, Bergen, Norway.

Evans, M. M. (2004) Land use and prey density changes in the Nakuru Wildlife Conservancy, Kenya: implications for cheetah conservation, Master of Science Thesis, University of Florida. Retrieved January 14, 2010, from www.carnivoreconservation.org/portal/index.php

ExperienceKenya (2010) *Cities and Towns.* Retrieved May 29, 2010, from www.experiencekenya.co.ke (click Cities and Towns).

Fan, X., Chou, J.-F., Guo, B.-R., and Shulski, M. D. (2004) A coupled simple climate model and its global analysis, *Theoretical and Applied Climatology,* 79, pp. 31–43.

FAO (Food and Agriculture Organization of the United Nations) (2008) *Maps.* Retrieved May 29, 2010, from www.fao.org/countryProfiles (select country, then click More maps).

Fearon, J. D. (1994) Domestic political audiences and the escalation of international disputes, *American Political Science Review,* 88, pp. 577–592.

Feinstein, A. H. and Cannon, H. M. (2001) Fidelity, verifiability, and validity of simulation: constructs for evaluation, *Developments in Business Simulation and Experiential Learning,* 28, pp. 57–67.

Foran, D. R., Crooks, K. R., and Minta, S. C. (1997) Species identification from scat: an unambiguous genetic method, *Wildlife Society Bulletin,* 25 (4), pp. 835–839.

Fox, H. E., Christian, C., Nordby, J. C., Pergams, O. R., Peterson, G. D., and Pyke, C. R. (2006) Perceived barriers to integrating social science and conservation, *Conservation Biology,* 20 (6), pp. 1817–1820.

Frankham, R., Ballou, J. D., and Briscoe, D. A. (2002) *Introduction to Conservation Genetics,* Cambridge University Press, Cambridge.

GBIF (2010) Data portal. Retrieved January 6, 2010, from http://data.gbif.org

Gibson, C. C. (1999) *Politicians and Poachers,* Cambridge University Press, Cambridge.

Good, P. I. (2005) *Permutation, Parametric, and Bootstrap Tests of Hypotheses,* Third Edition, Springer Series in Statistics XVI, Springer-Verlag, New York.

GOSIC (2010) GCOS essential climate variables (ECV) data access matrix, Global Observing Systems Information Center, maintained by the US National Oceanic and Atmospheric Administration. Retrieved May 29, 2010, from http://gosic.org/ios/MATRICES/ECV/ECV-matrix.htm

Green, E. D. (2005) Ethnicity and the politics of land tenure reform in Central Uganda, *Working Paper Series,* No. 05-58, Development Studies Institute, London School of Economics and Political Science, London.

Grimm, V. and Railsback, S. F. (2005) *Individual-Based Modeling and Ecology,* Princeton University Press, Princeton, New Jersey.

Gros, P. M. (1998) Status of the cheetah *Acinonyx jubatus* in Kenya: a field-interview assessment, *Biological Conservation,* 85, pp. 137–149.

—— (1999) Status and habitat preferences of Uganda cheetahs: an attempt to predict carnivore occurrence based on vegetation structure, *Biodiversity and Conservation,* 8, pp. 1561–1583.

—— (2002) The status and conservation of the cheetah *Acinonyx jubatus* in Tanzania, *Biological Conservation,* 106, pp. 177–185.

Haas, T. C. (1992) A Bayes network model of district ranger decision making, *Artificial Intelligence Applications,* 6 (3) pp. 72–88.

—— (1995) Local prediction of a spatio-temporal process with an application to wet sulfate deposition, *Journal of the American Statistical Association,* 90 (432), pp. 1189–1199.

—— (2001) A web-based system for public-private sector collaborative ecosystem management, *Stochastic Environmental Research and Risk Assessment,* 15 (2), pp. 101–131.

_____ (2002) New systems for modeling, estimating, and predicting a multivariate spatio-temporal process, *Environmetrics,* 13 (4), pp. 311–332.

_____ (2004) Ecosystem management via interacting models of political and ecological processes, *Animal Biodiversity and Conservation,* 27 (1), pp. 231–245. Retrieved May 29, 2010, from www.ben.cat/ABC (click Volums > Volum 27.1).

_____ (2008a) How to find the most practical ecosystem management plan, in *Environmental Economics and Investment Assessment II,* Proceedings of the Conference held in Cadiz, Spain, May 28–30, K. Aravossis, C. A. Brebbia, and N. Gomez (Eds.), Wessex Institute of Technology Press, Southampton, pp. 241–252.

_____ (2008b) East African Cheetah management via interacting political and ecological process models, in *Conserving and Valuing Ecosystem Services and Biodiversity,* K. N. Ninan (Ed.), Earthscan Publishers, London, pp. 221–260.

Haas, T. C. Mowrer, H. T., and Shepperd, W. D. (1994) Modeling Aspen stand growth with a temporal Bayes network, *Artificial Intelligence Applications,* 8 (1), pp. 15–28.

Hall, P. and Titterington, D. M. (1989) The effect of simulation order on level accuracy and power of Monte Carlo tests, *Journal of the Royal Statistical Society,* Series B, 51 (3), pp. 459–467.

Hand, D. J. (1997) *Construction and Assessment of Classification Rules,* John Wiley & Sons, Ltd, Chichester.

Handcock, M. S., Huovilainen, S. M., and Rendall, M. S. (2000) Combining registration-system and survey data to estimate birth probabilities, *Demography,* 37 (2), pp. 187–192.

Harrison, N. E. (2006) *Complexity in World Politics,* SUNY Series in Global Politics, State University of New York Press, Albany, New York.

Hauser, C. E. and Possingham, H. P. (2008) Experimental or precautionary? Adaptive management over a range of time horizons, *Journal of Applied Ecology,* 45, pp. 72–81.

Hayward, M. W., O'Brien, J., and Kerley, G. I. H. (2007) Carrying capacity of large African predators: predictions and tests, *Biological Conservation,* 139, pp. 219–229.

Healy, R. G. and Ascher, W. (1995) Knowledge in the policy process: incorporating new environmental information in natural resources policy making, *Policy Sciences,* 28, pp. 1–19.

Hedström, P. (2005) *Dissecting the Social: On the Principles of Analytical Sociology,* Cambridge University Press, Cambridge.

Helton, J. C. and Davis, F. J. (2000) Sampling-based methods, in *Sensitivity Analysis,* A. Saltelli, K. Chan, and E. M. Scott (Eds.), John Wiley & Sons, Inc., New York.

Henrion, M. (1988) Propagating uncertainty by logic sampling in Bayes' networks, in *Uncertainty in Artificial Intelligence 2,* J. F. Lemmer and L. N. Kanal (Eds.), North-Holland, Amsterdam.

Hilmer, C. E. and Holt, M. T. (2000) A comparison of resampling techniques when parameters are on a boundary: the bootstrap, subsample bootstrap, and subsample jackknife, Selected paper given at the *American Agricultural Economics Association Annual Meeting,* Tampa, Florida, July 30–August 2. Retrieved May 29, 2010, from http://purl.umn.edu/21810

Hoffmann, M., Kelley, H., and Evans, T. (2002) Simulating Land-Cover Change in South-Central Indiana: an Agent-Based Model of Deforestation and Afforestation, in M. A. Janssen (Ed.), *Complexity and Ecosystem Management: the Theory and Practice of Multi-Agent Systems,* Edward Elgar Publishing, New York, pp. 218–247.

Honolulu Zoo (2008) Cheetah. Retrieved January 12, 2010, from www.honoluluzoo.org/cheetah.htm

Hooke, R. and Jeeves, T. A. (1961) Direct search solution of numerical and statistical problems, *Journal of the ACM,* 8, pp. 212–229.

Hopcroft, R. L. (1998) The importance of the local: rural institutions and economic change in preindustrial England, in *The New Institutionalism in Sociology,* M. C. Brinton and V. Nee (Eds.), Russell Sage Foundation, New York, republished in 2001 by Stanford University Press, Stanford, California.

Hudson, V. M. (1983) The external predisposition component of a model of foreign policy behavior, Ph.D. Dissertation, Ohio State University, Columbus, Ohio.

Hurn, A. S., Lindsay, K. A., and Martin, V. L. (2003) On the efficacy of simulated maximum likelihood for estimating the parameters of stochastic differential equations, *Journal of Time Series Analysis,* 24 (1), pp. 45–63.

Hutto, R. L., Pletschet, S. M., and Hendricks, P. (1986) A fixed-radius point count method for nonbreeding and breeding season use, *The Auk,* 103 (July), pp. 593–602.

Iacus, S. M. (2008) *Simulation and Inference for Stochastic Differential Equations,* Springer-Verlag, New York.

IUCN (2009) Balaenoptera musculus, The IUCN Red List of Threatened Species. Retrieved December 31, 2009, from www.iucnredlist.org/apps/redlist/details/2477/0

Jank, W. (2006) Efficient simulated likelihood with an application to online retailing, *Statistics and Computing,* 16 (2), pp. 111–124.

Jank, W. and Booth, J. G. (2003) Efficiency of Monte Carlo EM and simulated maximum likelihood in two-stage hierarchical models, *Journal of Computational and Graphical Statistics,* 12, pp. 214–229.

Janssen, M. A., Anderies, J. M., Smith, M. S., and Walker, B. H. (2002) Implications of Spatial Heterogeneity of Grazing Pressure on the Resilience of Rangelands, in M. A. Janssen (Ed.), *Complexity and Ecosystem Management: the Theory and Practice of Multi-Agent Systems,* Edward Elgar Publishing, New York, pp. 103–123.

Johnson, J. H. (1994) Representation, knowledge elicitation and mathematical science, in *Artificial Intelligence in Mathematics,* J. H. Johnson, S. McKee, and A. Vella (Eds.), Clarendon Press, Oxford, pp. 313–328.

Jones, S. (1999) From meta-narratives to flexible frameworks: an actor level analysis of land degradation in highland Tanzania, *Global Environmental Change,* 9, pp. 211–219.

Jost, J. T. and Sidanius, J. (Eds.) (2004) *Political Psychology: Key Readings,* Psychology Press (Taylor & Francis), Abingdon.

Kaplan, D. (2009) *Structural Equation Modeling: Foundations and Extensions,* Second Edition, Sage, Los Angeles, California.

Karanth, K. U. and Nichols, J. D. (1998) Estimation of tiger densities in India using photographic captures and recaptures, *Ecology,* 79 (8), pp. 2852–2862.

Kaupe A. F. Jr., (1963) Algorithm 178: direct search, *Communications of the ACM,* 6, p. 313.

Kassimir, R. (1998) Uganda: the Catholic Church and state reconstruction, in *The African State at a Critical Juncture,* L. A. Villalón and P. A. Huxtable (Eds.), Lynne Rienner, Boulder, Colorado, pp. 233–254.

Keller, J. W. (2005) Constraint respecters, constraint challengers, and crisis decision making in democracies: a case study analysis of Kennedy versus Reagan, *Political Psychology,* 26 (6), pp. 835–867.

Kelly, M. J. and Durant, S. M. (2000) Viability of the Serengeti cheetah population, *Conservation Biology,* 14 (3), pp. 786–797.

Kiiveri, H., Speed, T. P., and Carlin, J. B. (1984) Recursive causal models, *Journal of the Australian Mathematical Society,* Series A, 36, pp. 30–52.

Kingdon, J. (1977) *East African Mammals: An Atlas of Evolution in Africa,* Academic Press, London.

Klerer, M. (1991) *Design of Very High-Level Computer Languages,* Second Edition, McGraw-Hill, New York.

Kloeden, P. E. and Platen, E. (1995) *Numerical Solution of Stochastic Differential Equations,* Second Corrected Printing, Springer-Verlag, New York.

Koopmans, L. H. (1969) Some simple singular and mixed probability distributions, *The American Mathematical Monthly,* 76 (3), pp. 297–299.

Koster, J. T. A. (1996) Markov properties of nonrecursive causal models, *The Annals of Statistics,* 24 (5), pp. 2148–2177.

_____ (1997a) Gibbs and Markov properties of graphs, *Annals of Mathematics and Artificial Intelligence,* 21 (1), pp. 13–26.

_____ (1997b) On the validity of the Markov interpretation of path diagrams of linear structural equations systems with correlated errors, *Erasmus University Technical Report EUR/FSW/97.03.01,* Erasmus University, Rotterdam.

Krosnick, J. and Chiang, I. (2009) *Explorations in Political Psychology,* Psychology Press (Taylor & Francis), Abingdon.

Laliberte, A. S. and Ripple, W. (2003) Automated wildlife counts from remotely sensed imagery, *Wildlife Society Bulletin,* 31 (2), pp. 362–371.

Lande, R., Engen, S., and Saether, B.-E. (2003) *Stochastic Population Dynamics in Ecology and Conservation,* Oxford University Press, Oxford.

Leeds, B. (1999) Domestic political institutions, credible commitments, and international cooperation, *American Journal of Political Science,* 43, pp. 979–1002.

Leitão, H. C. and Schiozer, D. J. (1999) A new automated history matching algorithm improved by parallel computing, *Society of Petroleum Engineers (SPE) Latin American and Caribbean Petroleum Engineering Conference,* Caracas, Venezuela, April 21–23. Retrieved January 12, 2010, from www.onepetro.org (SPE manuscript 53977).

Leng, R. J. (1999) *Behavioral Correlates of War, 1816–1979* (Computer File), 3rd Release, Middlebury College, Middlebury, Vermont, 1993, Study Number 8606 from the Inter-University Consortium for Political and Social Research (ICPSR), Ann Arbor, Michigan. Retrieved January 12, 2010, from www.icpsr.umich.edu

Leng, R. J. and Singer, J. D. (1988) Militarized interstate crises: the BCOW typology and its applications, *International Studies Quarterly,* 32, pp. 155–173.

Lewis, O. A. (2007) Review of *Complexity in World Politics* by Harrison, N. E. (2006), State University of New York Press, Albany, New York, USA, *Journal of Artificial Societies and Social Simulation,* 10 (4). Retrieved May 29, 2010, from http://jasss.soc.surrey.ac.uk (click Previous issues).

Lewis, R. M., Torczon, V., and Trosset, M. W. (2000) Direct search methods: then and now, *ICASE Report No. 2000-26,* NASA Langley Research Center, Hampton, Virginia.

Li, X. and Dang, S. (2007) Tactical battlefield entities simulation model based on multi-agent interactions, *Computational Science - ICCS 2007, Seventh International Conference,* Beijing China, May 27–30, *Proceedings, Part IV,* Lecture Notes in Computer Science 4490, Springer-Verlag, New York, pp. 121–128.

Lindblom, C. (1980) *The Policymaking Process,* Prentice Hall, New York.

Lindsay, B. (1994) Efficiency versus robustness: the case for minimum Hellinger distance and related methods, *The Annals of Statistics,* 22, pp. 1081–1114.

Liu, J., Dietz, T., Carpenter, S., Alberti, M., Folke, C., Moran, E., Pell, A., Deadman, P., Kratz, T., Lubchenco, J., Ostrom, E., Ouyang, Z., Provencher, W., Redman, C., Schneider, S., and Taylor, W. (2007) Complexity of coupled human and natural systems, *Science,* 317 (5844), pp. 1513–1516.

Long, N. and van der Ploeg, J. D. (1994) Heterogeneity, actor and structure: towards a reconstitution of the concept of structure, in *Rethinking Social Development: Theory, Research and Practice,* D. Booth (Ed.), Longman Scientific and Technical, Harlow, ch. 3.

Long, R. A, Donovan, T. M., MacKay, P., Zielinski, W. J., and Buzas, J. S. (2007) Effectiveness of scat detection dogs for detecting forest carnivores, *Journal of Wildlife Management,* 71 (6), pp. 2007–2017.

Lubinsky, D. J. (1990) Integrating statistical theory with statistical databases, *Annals of Mathematics and Artificial Intelligence,* 2, pp. 245–259.

Lucieer, A. (2010) *Research.* Retrieved January 13, 2010, from www.lucieer.net/research/heard.html

MacDonald, D. (2001) *The New Encyclopedia of Mammals,* Oxford University Press, Oxford.

Maddox, T. M. (2003) The ecology of cheetahs and other large carnivores in a pastoralist-dominated buffer zone, Ph.D. Thesis, University College, London and Institute of Zoology, London. Retrieved January 14, 2010, from www.carnivoreconservation.org/portal/index.php

Majda, A. Jl, Franzke, C., and Crommelin, D. (2009) Normal forms for reduced stochastic climate models, *Proceedings of the National Academy of Sciences,* 106 (10), pp. 3649–3653.

Manninen, T., Linne, M.-L., and Ruohonen, K. (2006) Developing Itô stochastic differential equation models for neuronal signal transduction pathways, *Computational Biology and Chemistry,* 30, pp. 280–291.

Manson, S. M. (2002) Validation and Verification of Multi-Agent Systems, in M. A. Janssen (Ed.), *Complexity and Ecosystem Management: the Theory and Practice of Multi-Agent Systems,* Edward Elgar Publishing, New York, pp. 63–74.

Marchand, E., Clément, F., Roberts, J. E., and Pépin, G. (2008) Deterministic sensitivity analysis for a model for flow in porous media, *Advances in Water Resources,* 31, pp. 1025–1037.

Mariano, R., Schuermann, T., and Weeks, M. J. (Eds.) (2000) *Simulation-Based Inference in Econometrics,* Cambridge University Press, Cambridge.

Marnewick, K., Funston, P. J., and Karanth, K. U. (2008) Evaluating camera trapping as a method for estimating cheetah abundance in ranching areas, *South African Journal of Wildlife Research,* 38 (1), pp. 59–65.

Matsubayashi, H., Lagan, P., Majalap, N., Tangah, J., Sukor, J. R. A., and Kitayama, K. (2007) Importance of natural licks for the mammals in Bornean inland tropical rain forests, *Ecological Research,* 22, pp. 742–748.

Mbugua, S. W. (1986) Monitoring livestock and wildlife in Kenya, in *Range Development and Research in Kenya,* Winrock International Institute for Agricultural Development, Morrilton, Arkansas.

Meadows, D. H., Meadows, D. L., Randers, J., and Behrens, W. W. (1972) *The Limits to Growth,* Potomac Associates–Universe Books, New York.

Metcalfe, S. and Kepe, T. (2008) Your elephant on our land, *The Journal of Environment and Development,* 17 (2), pp. 99–117.

Miles, S. B. (2000) Towards policy relevant environmental modeling: contextual validity and pragmatic models, *United States Geological Survey Open-File Report 00-401.*

Moir, W. H. and Block, W. M. (2001) Adaptive management on public lands in the United States: commitment or rhetoric? *Environmental Management,* 28 (2), pp. 141–148.

Mori, M. and Butterworth, D. S. (2004) Consideration of multispecies interactions in the Antarctic: a preliminary model of the minke whale – blue whale – krill interaction, *African Journal of Marine Science,* 26, pp. 245–259.

Morris, M. D. (1991) Factorial sampling plans for preliminary computational experiments, *Technometrics,* 33, pp. 161–174.

Murphy, J. M., Sexton, D. M. H., Barnett, D. N., Jones, G. S., Webb, M. J., Collins, M., and Stainforth, D. A. (2004) Quantification of modelling uncertainties in a large ensemble of climate change simulations, *Nature,* 430 (12 August), pp. 768–772.

Murray, S. K. and Cowden, J. A. (1999) The role of 'enemy images' and ideology in elite belief systems, *International Studies Quarterly,* 43, pp. 455–481.

Naveh, I. and Sun, R. (2006) Simulating a simple case of organizational decision making, in *Cognition and Multi-Agent Interaction: From Cognitive Modeling to Social Simulation,* R. Sun (Ed.), Cambridge University Press, Cambridge.

Neal, R. M. (1996) *Bayesian Learning for Neural Networks,* Springer-Verlag, New York.

New York Times (2007) Japan's whaling obsession, *The New York Times,* editorial published on April 1. Retrieved January 1, 2010, from www.nytimes.com

Nilsson, D. and Lauritzen, S. L. (2000) Evaluating influence diagrams using LIMIDs, in *Proceedings of the Sixteenth Conference on Uncertainty in Artificial Intelligence,* C. Boutilier and M. Goldszmidt (Eds.) Morgan Kaufmann, San Francisco, pp. 436–445.

North, D. C. (1990) *Institutions, Institutional Change, and Economic Performance,* Cambridge University Press, Cambridge.

——— (1998) Economic performance through time, in *The New Institutionalism in Sociology,* M. C. Brinton and V. Nee (Eds.), Russell Sage Foundation, New York, republished in 2001 by Stanford University Press, Stanford, California.

Nsimbe, J. V. (2008) Jobless find hope in Dubai, *The Observer,* August 13. Retrieved May 4, 2010, from www.observer.ug

O'Connell-Rodwell, C. E., Rodwell, T., Matthew, R. and Hart, L. A. (2000) Living with the modern conservation paradigm: can agricultural communities co-exist with elephants? A five-year case study in east Caprivi, Namibia, *Biological Conservation,* 93, pp. 381–391.

Otis, D. L., Burnham, K. P., White, G. C., and Anderson, D. R. (1978) Statistical inference from capture data on closed animal populations, *Wildlife Monographs,* 62, pp. 1–135.

Oucho, J. O. (2002) *Undercurrents of Ethnic Conflict in Kenya,* Brill, Leiden.

Page, S. E. (2008) Review of *Social Simulation: Technologies, Advances and New Discoveries,* B. Edmonds, C. Hernandez, and K. G. Troitzsch (Eds.) (2007) Information Science Reference, Hershey, Pennsylvania, USA, *Journal of Artificial Societies and Social Simulation,* 11 (2). Retrieved January 12, 2010, from http://jasss.soc.surrey.ac.uk

Pak, R. J. (1996) Minimum Hellinger distance estimation in simple linear regression models: distribution and efficiency, *Statistics and Probability Letters,* 26 (3), pp. 263–269.

Paolella, M. (2006) *Fundamental Probability: A Computational Approach,* John Wiley & Sons, Inc., Hoboken, New Jersey.

Park, S. and Gupta, S. (2009) Simulated maximum likelihood estimator for the random coefficient logit model using aggregate data, *Journal of Marketing Research,* 46 (4), pp. 531–542.

Parma, A. M. and the NCEAS Working Group on Population Management (1998) What can adaptive management do for our fish, forests, food and biodiversity? *Integrative Biology,* 1, pp. 16–26.

Partell, P. J. and Palmer, G. (1999) Audience costs and interstate crises: an empirical assessment of Fearon's model of dispute outcomes, *International Studies Quarterly,* 43, pp. 389–405.

Patra, R. K., Mandal, A., and Basu, A. (2008) Minimum Hellinger distance estimation with inlier modification, *Sankhyā,* 70-B, Part 2, pp. 310–322.

Pearl, J. (1988) *Probabilistic Reasoning in Intelligent Systems,* Morgan Kaufmann, San Mateo, California.

Peden, D. G. (1984) Livestock and wildlife population inventories by district in Kenya 1977–1983, *Technical Report 102,* Kenya Rangeland Ecological Monitoring Unit, Nairobi, Kenya.

Pelkey, N. W., Stoner, C. J., and Caro, T. M. (2000) Vegetation in Tanzania: assessing long term trends and effects of protection using satellite imagery, *Biological Conservation,* 94, pp. 297–309.

Phan, D. and Amblard, F. (Eds.) (2007) *Agent-Based Modelling and Simulation in the Social and Human Sciences,* The Bardwell Press, Oxford.

Politis, D. N. and Romano, J. P. (1994) Large sample confidence regions based on subsamples under minimal assumptions, *The Annals of Statistics,* 22 (4), pp. 2031–2050.

Punt, A. E. (1998) A full description of the standard baleen II model and some variants thereof, Paper SC/50/AWMP1 presented to the IWC Scientific Committee, May, available from the IWC Secretariat, The Red House, 135 Station Road, Impington, Cambridge CB4 9NP, UK.

Ramanujan, K. (2010) New tools for conservation, *NASA Earth Observatory*. Retrieved January 13, 2010, from http://earthobservatory.nasa.gov/Features/Conservation/printall.php

Read, A. J., Halpin, P. N., Crowder, L. B., Best, B. D., and Fujioka, E. (Eds.) (2010) *OBIS-SEAMAP: Mapping Marine Mammals, Birds, and Turtles.* Retrieved January 5, 2010, from http://seamap.env.duke.edu

Reise (2004) *Tanzania, Rwanda, Burundi 1:1 200 000,* World Mapping Project, Reise Know-How Verlag, Bielefeld. Retrieved January 12, 2010, from www.reise-know-how.de

Renaud, P. S. A. (1989) *Applied Political Economic Modelling,* Studies in Contemporary Economics, Springer-Verlag, New York.

Rosenbaum, P. R. (1986) Dropping out of high school in the United States: an observational study, *Journal of Educational Statistics,* 11 (3), pp. 207–224.

—— (1995) Discussion of 'Causal Diagrams for Empirical Research' by J. Pearl, *Biometrika,* 82 (4), pp. 698–699.

—— (2002) *Observational Studies,* Second Edition, Springer-Verlag, New York.

—— (2010) Design sensitivity and efficiency in observational studies, *Journal of the American Statistical Association,* 105 (490), pp. 692–702.

Rowcliffe, J. M., Field, J., Turvey, S. T., and Carbone, C. (2008) Estimating animal density using camera traps without the need for individual recognition, *Journal of Applied Ecology,* 45, pp. 1228–1236.

Rumelhart, D. E., Smolensky, P., McClelland, J. L., and Hinton, G. E. (1986) Schemata and sequential thought processes in parallel distributed processing models, in *Parallel Distributed Processing: Exploration in the Microstructure of Cognition, Volume 2,* D. E. Rumelhart, J. L. McClelland, and the PDP Research Group (Eds.), The MIT Press, Cambridge, Massachusetts, pp. 7–57.

Saadi, A. and Sahnoun, Z. (2006) Towards intentional agents to manipulate belief, desire and commitment degrees, *Proceedings of the Fourth ACS/IEEE International Conference on Computer Systems and Applications (AICCSA-06),* March 8–11, Dubai-Sharjah, United Arab Emirates, pp. 515–520.

Sabatier, P. A. and Jenkins-Smith, H. (1993) *Policy Change and Learning: An Advocacy Coalition Approach,* Westview Press, Boulder, Colorado.

Saltelli, A. (2000) What is sensitivity analysis? in *Sensitivity Analysis,* A. Saltelli, K. Chan, and E. M. Scott (Eds.), John Wiley & Sons, Inc., New York.

Sasamal, S. K., Chaudhury, S. B., Samal, R. N., and Pattanaik, A. K. (2008) QuickBird spots flamingos off Nalabana island, Chilika lake, India, *International Journal of Remote Sensing,* 29 (16), pp. 4865–4870.

Schrodt, P. A. (1995) Event data in foreign policy analysis, in *Foreign Policy Analysis,* L. Neack, J. A. K. Hey, and P. J. Haney (Eds.), Prentice Hall, Englewood Cliffs, New Jersey, pp. 145–166.

ScienceDaily (2009) Norway, Japan prop up whaling industry with taxpayer money, *ScienceDaily,* June 18. Retrieved December 31, 2009, from www.sciencedaily.com/releases/2009/06/090619082131.htm

Sears, D. O., Huddy, L., and Jervis, R. (2003) *Oxford Handbook of Political Psychology,* Oxford University Press, Oxford.

Seber, G. A. F. (1977) *Linear Regression Analysis,* John Wiley & Sons, Inc., New York.

Sedgewick, R. and Wayne, K. (2007) *Introduction to Programming in JAVA: An Interdisciplinary Approach,* Addison-Wesley, New York.

Service, S. *et al.* (2006) Magnitude and distribution of linkage disequilibrium in population isolates and implications for genome-wide association studies, *Nature Genetics,* 38 (5), pp. 556–560.

Shao, J. and Tu, D. (1995) *The Jackknife and Bootstrap,* Springer-Verlag, New York.

Shell, K. M. and Somerville, R. C. J. (2005) A generalized energy balance climate model with parameterized dynamics and diabatic heating, *Journal of Climate,* 18, pp. 1753–1772.

Silveira, L., Jácomo, A. T. A., and Diniz-Filho, J. A. F. (2003) Camera trap, line transect census, and track surveys: a comparative evaluation, *Biological Conservation,* 114 (3), pp. 351–355.

Smith, B. J. (2007) boa: an **R** package for MCMC output convergence assessment and posterior inference, *Journal of Statistical Software,* 21 (11), 37 pages. Retrieved May 27, 2010, from www.jstatsoft.org

Sokolov, A. P., Stone, P. H., Forest, C. E., Prinn, R., Sarofim, M. C., Webster, M., Paltsev, S., and Schlosser, C. A. (2009) Probabilistic forecast for twenty-first-century climate based on uncertainties in emissions (without policy) and climate parameters, *Journal of Climate,* October 1, pp. 5175–5204.

Spanos, A. (1995) On theory testing in econometrics: modeling with non-experimental data, *Journal of Econometrics,* 67, pp. 189–226.

Spring, D. A. and Kennedy, J. O. S. (2005) Existence value and optimal timber-wildlife management in a flammable multistand forest, *Ecological Economics,* 55, pp. 365–379.

Stephens, P. A., Zaumyslova, O. Yu., Miquelle, D. G., Myslenkov, A. I., and Hayward, G. D. (2006) Estimating population density from indirect sign: track counts and the Formozov-Malyshev-Pereleshin formula, *Animal Conservation,* 9 (3), pp. 339–348.

Stoett, P. J. (1997) *The International Politics of Whaling,* University of British Columbia Press, Vancouver.

Sun, R. (Ed.) (2006) *Cognition and Multi-Agent Interaction: From Cognitive Modeling to Social Simulation,* Cambridge University Press, Cambridge.

Tageo (2010a) *United Republic of Tanzania (TZ).* Retrieved May 29, 2010, from www.tageo.com/index-e-tz-cities-TZ.htm

—— (2010b) *Uganda (UG).* Retrieved May 29, 2010, from www.tageo.com/index-e-ug-cities-UG.htm

Taha, H. A. (1975) *Integer Programming,* Academic Press, New York.

Takada, T. (2009) Simulated minimum Hellinger distance estimation of stochastic volatility models, *Computational Statistics and Data Analysis,* 53 (6), pp. 2390–2403.

Tamura, R. N. and Boos, D. D. (1986) Minimum Hellinger distance estimation for multivariate location and covariance, *Journal of the American Statistical Association,* 81 (393), pp. 223–229.

TEEB (The Economics of Ecosystems and Biodiversity) (2010) Research for the Environment. Retrieved August 9, 2010, from www.ufz.de/index.php?en=17633

Thomas, L., Buckland, S. T., Burnham, K. P., Anderson, D. R., Laake, J. L., Borchers, D. L., and Strindberg, S. (2002) Distance Sampling, in A. H. El-Shaarawi and W. W. Piegorsch (Eds.), *Encyclopedia of Environmetrics,* John Wiley & Sons, Ltd, Chichester, pp. 544–552.

Thompson, J. R. and Tapia, R. A. (1990) *Nonparametric Function Estimation, Modeling and Simulation,* Society for Industrial and Applied Mathematics, Philadelphia, Pennsylvania.

Thompson, S. K. (1992) *Sampling,* John Wiley & Sons, Inc., New York.

Throup, D. and Hornsby, C. (1998) *Multi-Party Politics in Kenya,* Ohio University Press, Athens, Ohio.

Tilman, D. (1982) *Resource Competition and Community Structure,* Princeton University Press, Princeton, New Jersey.

TMAP (2008) *Tanzania Mammal Atlas Project (TMAP),* part of the Tanzania Mammal Conservation Program maintained by the Tanzania Wildlife Research Institute, Arusha, Tanzania. Retrieved January 12, 2010, from www.tanzaniamammals.org

Tomlin, F. K. and Smith, I. B. (1969) Remark on algorithm 178 [E4] direct search, *Communications of the ACM,* 12, pp. 637–638.

UFZ (Helmholtz Centre for Environmental Research) (2008) Biodiversity as a natural resource, Press Release May 21, Helmholtz Centre for Environmental Research. Retrieved August 9, 2010, from www.ufz.de/index.php?en=16772

Uleman, J. S. (1996) When do unconscious goals cloud our minds? In *Ruminative Thoughts,* R. S. Wyer, Jr. (Ed.), Taylor & Francis, New York, pp. 165–176.

UN (United Nations) (2008a) General map of Tanzania. Retrieved May 29, 2010, from www.un.org (click Documents > Maps and Geographic Information > General Maps and then select Tanzania).

—— (2008b) General map of Uganda. Retrieved May 29, 2010, from www.un.org (click Documents > Maps and Geographic Information > General Maps and then select Uganda).

_____ (2010) The World. Retrieved February 15, 2010, from www.un.org/Depts/ Cartographic/map/profile/world.pdf

USAID (United States Agency for International Development) (2005) *Biodiversity Conservation: A Guide for USAID Staff and Partners,* United States Agency for International Development. Retrieved August 9, 2010, from http://pdf.usaid .gov/pdf_docs/Pnade258.pdf

US Army (2007) Request for information (RFI) 2007-1276: agent-based modeling of irregular warfare (ABMIW), Department of the Army, Alexandria, Virginia, USA. Retrieved May 29, 2010, from www.fbo.gov (click Opportunities > Advanced Search > Archived Documents, then enter the solicitation number: RFI-2007-1276).

USGS (United States Geological Survey) (1997) Recent highlights – natural resources, *Factsheet FS-010-97.* Retrieved August 9, 2010, from www.usgs.gov/themes/FS-010-97/

Uyaphi (2008) *Uganda Search Map.* Retrieved January 12, 2010, from www.uyaphi .com/search/map/uganda

Vermeulen, P. J. and De Jongh, D. C. J. (1977) Growth in a finite world – a comprehensive sensitivity analysis, *Automatica,* 13, pp. 77–84.

Vertzberger, Y. Y. I. (1990) *The World in Their Minds,* Stanford University Press, Stanford, California.

Wambua, C. (2008) Monitoring game, *Cheetah Conservation Fund – Kenya Newsletter,* 4, August. Retrieved January 17, 2010, from www.cheetah.org/ama/ orig/CCFKissue4-0808.pdf

Weber, D. C. and Skillings, J. H. (2000) *A First Course in the Design of Experiments: A Linear Models Approach,* CRC Press, Boca Raton, Florida.

Weible, C. M. (2007) An advocacy coalition framework approach to stakeholder analysis: understanding the political context of California marine protected area policy, *Journal of Public Administration Research and Theory,* 17 (1), pp. 95–117.

Wells, H., Strauss, E. G., Rutter, M. A., and Wells, P. H. (1998) Mate location, population growth and species extinction, *Biological Conservation,* 86, pp. 317–324.

Wikipedia (2009a) Blue whale. Retrieved August 27, 2009, from http://en .wikipedia.org/wiki/Blue_whale

_____ (2009b) Whaling in Iceland. Retrieved December 31, 2009, from http://en .wikipedia.org/wiki/Whaling_in_Iceland

_____ (2009c) Whaling in Norway. Retrieved December 31, 2009, from http://en .wikipedia.org/wiki/Whaling_in_Norway

_____ (2009d) International Whaling Commission (IWC). Retrieved January 1, 2010, from http://en.wikipedia.org/wiki/International_Whaling_ Commission

_____ (2010) Natural resource. Retrieved August 9, 2010, from http://en .wikipedia.org/wiki/Natural_resource.

World Bank (2008) *Country Classification.* Retrieved January 12, 2010, from www.worldbank.org (click Data and Research > Data Website > Data and Statistics > Country Classification).

_____ (2009) *The Worldwide Governance Indicators (WGI) Project.* Retrieved May 29, 2010, from http://info.worldbank.org/governance/wgi/index.asp (click Governance Indicators webpage and then select a country).

World Resources Institute (2005) *Population, Health and Human Well-Being Country Profiles*. Retrieved May 29, 2010, from http://earthtrends.wri.org

Yang, A., Abbass, H. A., and Sarker, R. (2005) WISDOM-II: a network centric model for warfare, *Knowledge-Based Intelligent Information and Engineering Systems Proceedings of the Ninth International Conference, KES 2005*, Lecture Notes in Computer Science, 3683, pp. 813–819, Springer-Verlag, New York.

Yasuda, M., Matsubayashi, H., Rustam, S. N., Sukor, J. R. A., and Bakar, S. A. (2007) Recent cat records by camera traps in peninsular Malaysia and Borneo, *CAT News,* 47, pp. 14–16.

Yodzis, P. (1994) Predator-prey theory and management of multispecies fisheries, *Ecological Applications,* 4 (1), pp. 51–58.

Zhou, S. and Griffiths, S. P. (2007) Estimating abundance from detection-nondetection data for randomly distributed or aggregated elusive populations, *Ecography,* 30, pp. 537–549.

Index

Improving Natural Resource Management: Ecological and Political Models   Timothy C. Haas
© 2011 John Wiley & Sons, Ltd

242   INDEX

STATISTICS IN PRACTICE

Human and Biological Sciences

Berger – Selection Bias and Covariate Imbalances in Randomized Clinical Trials
Berger and Wong - An Introduction to Optimal Designs for Social and
    Biomedical Research
Brown and Prescott - Applied Mixed Models in Medicine, Second Edition
Carstensen – Comparing Clinical Measurement Methods
Chevret (Ed) – Statistical Methods for Dose-Finding Experiments
Ellenberg, Fleming and DeMets – Data Monitoring Committees in Clinical Trials:
    A Practical Perspective
Hauschke, Steinijans & Pigeot – Bioequivalence Studies in Drug Development:
    Methods and Applications
Lawson, Browne and Vidal Rodeiro – Disease Mapping with WinBUGS and
    MLwiN
Lesaffre, Feine, Leroux & Declerck - Statistical and Methodological Aspects of
    Oral Health Research
Lui – Statistical Estimation of Epidemiological Risk
Marubini and Valsecchi - Analysing Survival Data from Clinical Trials and
    Observation Studies
Molenberghs and Kenward – Missing Data in Clinical Studies
O'Hagan, Buck, Daneshkhah, Eiser, Garthwaite, Jenkinson, Oakley & Rakow –
    Uncertain Judgements: Eliciting Expert's Probabilities
Parmigiani – Modeling in Medical Decision Making: A Bayesian Approach
Pintilie – Competing Risks: A Practical Perspective
Senn - Cross-over Trials in Clinical Research, Second Edition
Senn - Statistical Issues in Drug Development, Second Edition
Spiegelhalter, Abrams and Myles – Bayesian Approaches to Clinical Trials and
    Health-Care Evaluation
Walters - Quality of Life Outcomes in Clinical Trials and Health-Care Evaluation
Whitehead - Design and Analysis of Sequential Clinical Trials, Revised
    Second Edition
Whitehead – Meta-Analysis of Controlled Clinical Trials
Willan and Briggs – Statistical Analysis of Cost Effectiveness Data
Winkel and Zhang - Statistical Development of Quality in Medicine

Earth and Environmental Sciences

Buck, Cavanagh and Litton – Bayesian Approach to Interpreting Archaeological
    Data
Glasbey and Horgan – Image Analysis in the Biological Sciences
Haas – Improving Natural Resource Management: Ecological and Political
    Models

Helsel – Nondetects and Data Analysis: Statistics for Censored Environmental Data

Illian, Penttinen, Stoyan, H and Stoyan D–Statistical Analysis and Modelling of Spatial Point Patterns

McBride – Using Statistical Methods for Water Quality Management

Webster and Oliver – Geostatistics for Environmental Scientists, Second Edition

Wymer (Ed) – Statistical Framework for Recreational Water Quality Criteria and Monitoring

Industry, Commerce and Finance

Aitken - Statistics and the Evaluation of Evidence for Forensic Scientists, Second Edition

Balding - Weight-of-evidence for Forensic DNA Profiles

Brandimarte – Numerical Methods in Finance and Economics: A MATLAB-Based Introduction, Second Edition

Brandimarte and Zotteri – Introduction to Distribution Logistics

Chan - Simulation Techniques in Financial Risk Management

Coleman, Greenfield, Stewardson and Montgomery (Eds) – Statistical Practice in Business and Industry

Frisen (Ed) – Financial Surveillance

Fung and Hu – Statistical DNA Forensics

Gusti Ngurah Agung - Time Series Data Analysis Using EViews

Kenett (Eds) - Operational Risk Management: A Practical Approach to Intelligent Data Analysis

Jank and Shmueli (Ed.) – Statistical Methods in e-Commerce Research

Lehtonen and Pahkinen - Practical Methods for Design and Analysis of Complex Surveys, Second Edition

Ohser and Mücklich - Statistical Analysis of Microstructures in Materials Science

Pourret, Naim & Marcot (Eds) – Bayesian Networks: A Practical Guide to Applications

Taroni, Aitken, Garbolino and Biedermann - Bayesian Networks and Probabilistic Inference in Forensic Science

Taroni, Bozza, Biedermann, Garbolino and Aitken – Data Analysis in Forensic Science